Dedicated to Professor E. Heinz

PREFACE

These lecture notes have been written as an introduction to the characteristic theory for two-dimensional Monge–Ampère equations, a theory largely developed by H. Lewy and E. Heinz which has never been presented in book form. An exposition of the Heinz–Lewy theory requires auxiliary material which can essentially be found in various monographs, but which is presented here, in part because the focus is different, and also because these notes have an introductory character and are intended to be essentially self contained: Included are excerpts from the regularity theory for elliptic systems, the theory of pseudoanalytic functions and the theory of conformal mappings. References are usually scarce in the text; the sources are listed in the introduction.

These notes grew out of a seminar given in the Department of Mathematics of the University of Kentucky in Lexington during the Fall of 1988. I gratefully acknowledge the support of N.S.F. grant RII–8610671 and the Commonwealth of Kentucky through the Kentucky EPSCoR Program. It is a pleasure to thank Ronald Gariepy for inviting me to give the seminar, and I wish to express my sincere gratitude to all participants for their warm hospitality. Special thanks also to David Adams, Craig Evans and Neil Trudinger, whose interest encouraged me to actually write these notes. I am grateful for the corrections of George Paulik, and I appreciate the patience and dedication of Julie Hill in typing the bulk of the manuscript.

Iowa City, May 1990

TABLE OF CONTENTS

INTRODUCTION

An outline of the book, some historical remarks and the sources are given in this introductory section.

The present notes are mainly concerned with the "characteristic" theory for elliptic Monge–Ampère equations. This theory was largely developed by H. Lewy [L 3, 4] and E. Heinz [H 1–3, 5–7, 11, 12], motivated and based on the characteristic theory for hyperbolic equations as presented in Courant–Hilbert [CH] and Hadamard [HA 2] and on the "characteristic" theory for hyperbolic surfaces as developed by Darboux [DB] and F. Rellich [RE 2], who made the connection between the characteristic theory for differential equations and the surface theory.

Most noteworthy seems to be Appendix 4 to Chapter V of [CH] concerning the special role of the hyperbolic Monge–Ampère equation in the characteristic theory, namely the fact that the characteristic system of this fully nonlinear equation consists of only five equations instead of eight, a property that it shares with quasilinear equations. This fact was instrumental when Lewy founded the "characteristic" theory for elliptic Monge–Ampère equations.

Remarkable, and nowadays standard knowledge, are Appendices 1 and 2 to Chapter V of [CH] on Lewy's method [L 2] to accomplish the change from the hyperbolic to the elliptic case via a complex substitution.

In the "characteristic" theory for hyperbolic surfaces, asymptotic line parameters are constructed. A complex substitution (Rellich [R 2]) yields the "elliptic" case of convex surfaces, in which the "characteristic" variables are the conjugate isothermal parameters.

To be more precise, when introducing conjugate isothermal parameters x, y for a locally convex surface Σ, the second fundamental form

$$II_\Sigma = L\,du^2 + 2M\,du\,dv + N\,dv^2 \tag{1}$$

is reduced to

$$II_\Sigma = \Lambda(dx^2 + dy^2), \qquad \Lambda > 0. \tag{2}$$

The mapping $(u(x,y), v(x,y))$ satisfies a second order elliptic system, which can be written in the form

$$Lu = \Gamma^1_{11}\,|Du|^2 + \Gamma^1_{12}\,Du\cdot Dv + \Gamma^1_{22}\,|Du|^2, \tag{3}$$

$$Lv = \Gamma^2_{11}\,|Du|^2 + \Gamma^2_{12}\,Du\cdot Dv + \Gamma^2_{22}\,|Du|^2, \tag{4}$$

$$L = -\frac{1}{\sqrt{K}}\left[\frac{\partial}{\partial x}\left[\sqrt{K}\frac{\partial}{\partial x}\right] + \frac{\partial}{\partial y}\left[\sqrt{K}\frac{\partial}{\partial y}\right]\right]. \tag{5}$$

Here K is the Gauß curvature and the Γ_{ij}^{k}'s are the Christoffel symbols of the first fundamental form

(6) $$I_\Sigma = E\,du^2 + 2F\,du\,dv + G\,dv^2.$$

The fact that the coefficients of the system (3,4) depend only on the coefficients of the first fundamental form (6) and their derivatives is truly remarkable, a theorema egregium in the sense of Gauß.

The "Darboux system" (3,4) is the harmonic map system if K is constant. It can be derived at least formally by a complex substitution from the classical Darboux system for hyperbolic surfaces [DB].

The main thrust of the present notes is to present the Heinz–Lewy "characteristic" theory for elliptic Monge–Ampère equations. Consider the characteristic form

(7) $$ds^2 = (r+C)\,dx^2 + 2(s-B)\,dx\,dy + (t+A)\,dy^2$$

(8) $$= a\,dx^2 + 2b\,dx\,dy + c\,dy^2$$

associated with the elliptic Monge–Ampère equation

(9) $$Ar + 2Bs + Ct + (rt - s^2) = E,$$

or equivalently

(10) $$(r+C)(t+A) - (s-B)^2 = \Delta,$$

(11) $$\Delta = AC - B^2 + E > 0,$$

for a given solution $z = z(x,y)$. Here $p = z_x$, $q = z_y$, $r = z_{yy}$, $s = z_{xy}$, $t = z_{yy}$. New variables u, v are introduced such that

(12) $$ds^2 = \Lambda(du^2 + dv^2), \qquad \Lambda \neq 0.$$

We shall call the parameters u, v "characteristic", although not in the literal sense. Thus a conformal map with respect to the Riemannian metric ds^2 is constructed and the corresponding Beltrami system is

(13) $$x_u = \frac{b x_v + c y_v}{\sqrt{\Delta}},$$

(14) $$x_v = \frac{-b x_u - c y_u}{\sqrt{\Delta}}.$$

As far as the Monge–Ampère equation is concerned, one wishes to obtain information about the second derivatives, i.e., about the coefficients of the characteristic form. This means that the coefficients of the Beltrami system are "unknown". The difficulty is resolved by the observation (a theorema egregium of sorts) that the inverse mapping $(x(u,v), y(u,v))$, in addition to solving a Beltrami system, satisfies a quasilinear elliptic system of second order with quadratic growth in the gradient of the solution mapping of the form

$$Lx = h_1 |Dx|^2 + h_2 Dx \cdot Dy + h_3 |Dy|^2 + h_4 Dx \wedge Dy, \tag{15}$$

$$Ly = \tilde{h}_1 |Dx|^2 + \tilde{h}_2 Dx \cdot Dy + \tilde{h}_3 |Dy|^2 + \tilde{h}_4 Dx \wedge Dy, \tag{16}$$

$$L = -\frac{1}{\sqrt{\Delta}} \left[\frac{\partial}{\partial u} \left[\sqrt{\Delta} \frac{\partial}{\partial u} \right] + \frac{\partial}{\partial v} \left[\sqrt{\Delta} \frac{\partial}{\partial v} \right] \right]. \tag{17}$$

The coefficients $h_1, \ldots, \tilde{h}_4$ can be computed in terms of p, q and certain derivatives of A, B, C and Δ. The "characteristic" system (15, 16) reduces to

$$Lx = 0, \tag{18}$$

$$Ly = 0 \tag{19}$$

in the case of the simple Monge–Ampère equation

$$rt - s^2 = \Delta > 0. \tag{20}$$

This means that one should study diffeomorphic solutions of quasilinear elliptic second order systems. The corresponding conformality relations can be written in the form

$$\frac{t+A}{\sqrt{\Delta}} = \frac{|Dx|^2}{J(x,y)}, \tag{21}$$

$$-\frac{s-B}{\sqrt{\Delta}} = \frac{Dx \cdot Dy}{J(x,y)}, \tag{22}$$

$$\frac{r+C}{\sqrt{\Delta}} = \frac{|Dy|^2}{J(x,y)}, \tag{23}$$

$$J(x,y) = x_u y_v - x_v y_u, \tag{24}$$

and they can be used as a "dictionary", namely to translate the information obtained for the second order system, such as regularity or a priori estimates, into information for the Monge–Ampère equation.

It turns out that the regularity theory for Monge–Ampère equations can be presented much more directly by studying the Legendre–like variable transformation

(25) $$u = x,$$

(26) $$v = q$$

for the simple Monge–Ampère equation (20). This is of course motivated by the characteristic theory. The "characteristic" system (18,19) is replaced by the equation

(27) $$y_{uu} + (\Delta y_v)_v = 0.$$

This and the extension to general Monge–Ampère equations (9) are the contents of CHAPTER 3. The Campanato regularity technique is developed for Monge–Ampère equations, thus providing a scheme to prove sharp a priori estimates assuming the knowledge of bounds for the absolute values of the solution and its derivatives up to the second order. In order to continue paraphrasing Lewy's remarks from 1934 ([L3] loc. cit.) it seems, however, that the estimation of the second derivatives themselves requires much deeper insight. Purely local second derivative estimates can, at this point in time, only be shown via the characteristic theory which is the topic of Chapter 9. The Schauder technique could have been employed to yield the a priori estimates mentioned above (see Schulz [SZ3]), but, it seems, not the regularity results. The presentation of Chapter 3 is based on Schulz [SZ1–4] and Schulz–Liao [SL]. Historical references for the Legendre–like transformation (25,26) are Darboux [DB], Heinz [H5], Hartman–Wintner [HW1,3], Jörgens [JÖ1].

Chapter 3 requires the regularity theory for linear elliptic equations, in particular the Campanato technique. This is presented in CHAPTER 2 basedon Campanato [C3,4] and Giaquinta [GI].

Basic tools, in particular the concept of Hölder continuity, which are needed in Chapter 2 and later are presented in CHAPTER 1. The sources are: Campanato [C1,2], Evans–Gariepy [EG], Giaquinta [GI], Gilbarg–Trudinger [GT] and Heinz [H2].

Quasilinear elliptic second order systems are studied in CHAPTERS 2, 5 and 8. In Chapter 2 (Section 4), the regularity theory for univalent solutions is presented based on Schulz [SZ5]. In Chapters 5 and 8 diffeomorphic solutions of Heinz–Lewy type systems (15,16) are studied. The non–vanishing of the Jacobian is shown together with an a priori estimate from below. Chapter 5 deals with a special case which can be proved with the similarity principle. The reference is Heinz [H2]. Chapter 8 is about the general case without the similarity principle. The presentation is based on Heinz [H6,11] and Schulz [SZ5]. The general case via the similarity principle is not presented here. This topic would be an extension of Heinz [H12]. The fundamental historical reference is Lewy [L4], whose ideas are incorporated in the text, in particular in the proof of Proposition 8.1.2. Other references are Berg [BG] and Bers [BS3], who studied univalent solutions of linear systems.

Function theoretic tools which are needed in Chapter 5 are presented in CHAPTER 4. The main theorem is the similarity principle for pseudoanalytic functions by Bers and Vekua [BS 1, 2], [VE 1, 2], a $\bar{\partial}$-proof of which is presented, and a Harnack type inequality. The sources are: Bers [BS 2], Goursat [GO], Heinz [H 2] and Vekua [VE 2].

Tools needed in Chapter 8 are presented in CHAPTER 7. Function theoretic properties of elliptic equations are presented which cannot directly be derived from the similarity principle. The local behavior of functions satisfying elliptic differential inequalities is studied. The sources are: Hartman–Wintner [HW 2], Heinz [H 6] and Schulz [SZ 5].

Conformal mappings with respect to a Riemannian metric are studied in CHAPTER 6. The focus here is however somewhat different than in the standard literature about the Riemann mapping theorem. Our interest lies in the connections between uniformization and second order elliptic systems. The sources are: Ahlfors [AF], Heinz [H 3], Schiffer–Spencer [SHS], Schulz [SZ 6] and Vekua [VE 2]. Some results are used in Chapter 7, but the major applications are presented in Chapters 9 and 10, namely the connection between Monge–Ampère equations and quasilinear elliptic systems and the role of the Darboux system in the theory of convex surfaces.

CHAPTER 9 can be considered the core of the current notes. It is concerned with the characteristic theory for elliptic Monge–Ampère equations as outlined at the beginning of this introduction. Characteristic parameters are introduced by employing Chapter 6. The results of Chapters 2, 5, 8 on second order elliptic systems are translated into results for Monge–Ampère equations via the conformality relations (21, 22, 23). A priori estimates are thus derived for the absolute values of the second derivatives r, s, t of solutions of Monge–Ampère equations. The presentation is based on Heinz [H 3, 7] and Schulz [SZ 6].

Some geometric applications of Chapter 9 are discussed in CHAPTER 10, such as convex graphs of prescribed Gauß curvature and convex surfaces via the Darboux equation (a Monge–Ampère type equation which is different from the Darboux system). The main thrust of this chapter however is the investigation of locally convex surfaces without utilizing Monge–Ampère equations. Conjugate isothermal parameters x, y are introduced, and the Darboux system (3, 4) is derived as the "characteristic" system. A priori estimates are derived for the coefficients of the second fundamental form (1). This is based on Heinz [H 6] and Schulz [SZ 7].

Many interesting topics could not be covered in these lecture notes, most notably the special case of harmonic mappings, Jörgens's theorem for the equation $rt - s^2 = 1$ and applications to minimal surfaces. The investigation of the differential inequality $\alpha \le rt - s^2 \le \beta$ would have high priority for inclusion in an expanded version of these notes.

In addition to listing the sources at the end of the notes, the BIBLIOGRAPHY includes some related work, in particular two-dimensional Monge–Ampère equations and classical surface theory in three-space. The focus here is two-dimensional and work on multidimensional problems was not included.

The notation used is usually explained as it occurs. Because of the variety of the topics presented, we experimented with various notation such as complex, two–dimensional and multidimensional index notation. For most of the more subtle sections, it seemed necessary to employ the two–dimensional notation used in this introduction. There is a NOTATION INDEX on page 121, basic notation is explained on the following page.

BASIC NOTATION

$B_R = B_R(x)$ is the open ball or disc in $\mathbb{R}^n$ of radius R centered at x $(n \geq 2)$.
$\mathcal{N} = \mathcal{N}(x)$ is an open neighborhood of x, i.e., an open set containing x.
Ω denotes an open subset of $\mathbb{R}^n$; Ω is a domain if it is also connected.
$\Omega' \subset\subset \Omega$ means that the closure of Ω' is compact and contained in Ω.
e_i is the ith standart unit vector in $\mathbb{R}^n$.
$|x| = (\Sigma x_i^2)^{1/2}$ for a point $x = (x_1,\ldots,x_n)$ in $\mathbb{R}^n$.
$|\alpha| = \alpha_1 + \ldots + \alpha_n$ for a multiindex $\alpha = (\alpha_1,\ldots,\alpha_n)$ in $\mathbb{Z}^n$, $\alpha_i \geq 0$.
$x^\alpha = x_1^{\alpha_1} \cdot \ldots \cdot x_n^{\alpha_n}$.
$D_\alpha u = \partial u/\partial x_\alpha$, $D_{\alpha\beta} u = \partial^2 u/\partial x_\alpha \partial x_\beta, \ldots$ $(\alpha,\beta,\ldots = 1,\ldots,n)$.
$Du = (D_1 u,\ldots,D_n u)$ is the gradient of u.
$D^\alpha u = \partial^{|\alpha|}/\partial x_1^{\alpha_1} \ldots \partial x_n^{\alpha_n}$ for a multiindex $\alpha = (\alpha_1,\ldots,\alpha_n)$.
$C^k_{loc}(\Omega)$ is the set of functions having continuous derivatives of order $\leq k$ in Ω $(0 \leq k \leq \infty)$.
$C^k(\Omega)$ is the set of functions in $C^k_{loc}(\Omega)$ with finite C^k-norm

$$\|u\|_{C^k(\Omega)} = \sum_{\ell=1}^{k} |D^\ell u| = \sum_{\ell=1}^{k} \sup_{|\alpha|=\ell} \sup_{\Omega} |D^\alpha u| .$$

$C^k_0(\Omega)$ is the set of functions u in $C^k_{loc}(\Omega)$ with compact support in Ω, i.e., the closure of the set on which $u \neq 0$ is compact (and contained in Ω).
$C^{k,1}_{loc}(\Omega)$ is the set of functions in $C^k_{loc}(\Omega)$ whose k^{th} order derivatives are Lipschitz continuous in every $\Omega' \subset\subset \Omega$.
$L^p(\Omega)$ is the set of equivalence classes of measurable functions u on Ω which agree a. e. with finite L^p-norm

$$\|u\|_{L^p(\Omega)} = \left[\int_\Omega |u(x)|^p \, dx\right]^{1/p} .$$

Note: In practice we usually do not identify two L^p-functions which agree a. e. and work instead with the precise representative which is given by the Lebesgue differentiation theorem.
$L^p_{loc}(\Omega)$ is the set of functions in $L^p(\Omega')$ for every $\Omega' \subset\subset \Omega$.
$\sup_\Omega u$ denotes the essential supremum of u, i.e. the quantity $\operatorname{ess\,sup} u = \|u^+\|_{L^\infty(\Omega)}$.
$\operatorname{osc}_\Omega u$ is the essential oscillation of u in Ω, i. e. equal to $\operatorname{ess\,osc} u = \sup_\Omega u - \inf_\Omega u$.
$C = C(\ldots)$ denotes a constant which depends only on the quantities that are listed in parentheses. The letter C will denote various constants which may change from line to line. We choose constants to be ≥ 1, if possible. Generic constants that are < 1 are usually denoted by the lower case letter c.

Chapter 1. INTEGRAL CRITERIA FOR HÖLDER CONTINUITY

1.1. Sobolev functions

Let Ω be an open set in $\mathbb{R}^n$, $n \geq 2$, with a typical point denoted by $x = (x_1, \ldots, x_n)$. The Sobolev space $W^{k,p}(\Omega)$, $k = 0, 1, 2, \ldots$, $1 \leq p \leq \infty$, is the space of functions $u \in L^p(\Omega)$ with weak partial derivatives up to order k in $L^p(\Omega)$. Here, for any multiindex $\alpha = (\alpha_1, \ldots, \alpha_n)$ with $|\alpha| = \alpha_1 + \ldots + \alpha_n$, $v = D^\alpha u = \partial^{|\alpha|} u / \partial x_1^{\alpha_1} \ldots \partial x_n^{\alpha_n}$ is the αth weak derivative of u in Ω if

$$(1.1) \qquad \int_\Omega v \eta \, dx = (-1)^{|\alpha|} \int_\Omega u \, D^\alpha \eta \, dx$$

for all $\eta \in C_0^\infty(\Omega)$.

The function u belongs to $W^{k,p}_{loc}(\Omega)$ if $u \in W^{k,p}(\Omega')$ for each $\Omega' \subset\subset \Omega$. Vector valued Sobolev functions $u = (u^1, \ldots, u^N) \in W^{k,p}(\Omega, \mathbb{R}^N)$ are defined in the obvious way.

$W^{k,p}(\Omega)$ is a Banach space if equipped with the Sobolev norm

$$(1.2) \qquad \|u\|_{k,p} = \|u\|_{W^{k,p}(\Omega)} = \Big[\int_\Omega \sum_{|\alpha| \leq k} |D^\alpha u|^p \, dx\Big]^{1/p}$$

for $1 \leq p < \infty$, and, with $\sup$ denoting the essential supremum,

$$(1.3) \qquad \|u\|_{k,\infty} = \|u\|_{W^{k,\infty}(\Omega)} = \sup_\Omega \sum_{|\alpha| \leq k} |D^\alpha u| \, .$$

$W_0^{k,p}(\Omega)$ is the closure of $C_0^\infty(\Omega)$ in the space $W^{k,p}(\Omega)$.

Sobolev functions can be approximated by smooth functions on Ω: If $1 \leq p < \infty$, then a theorem of Meyers–Serrin [MS] states for arbitrary Ω that

$$(1.4) \qquad W^{k,p}(\Omega) = H^{k,p}(\Omega) ,$$

which is the completion of $\{u \in C^k_{loc}(\Omega) \mid \|u\|_{k,p} < \infty\}$ with respect to the norm $\|\cdot\|_{k,p}$. In the case $p = \infty$,

$$(1.5) \qquad W^{k,\infty}_{loc}(\Omega) = C^{k-1,1}_{loc}(\Omega) ,$$

the space of functions whose $(k-1)^{st}$ derivatives are locally Lipschitz continuous in Ω.

A Sobolev function of class $W^{1,p}_{loc}(\mathbb{R}^n)$ is precisely represented by a function u which is locally absolutely continuous on almost all lines, i.e., $u_i(t) = u(\ldots, x_{i-1}, t, x_i, \ldots)$ is absolutely continuous on compact subintervals of $\mathbb{R}$ for almost every $(x_1, \ldots, x_{i-1}, x_{i+1}, \ldots, x_n)$, and $Du_i \in L^p_{loc}(\mathbb{R}^n)$.

1.2. The Dirichlet growth theorem

Let $B_R = B_R(x)$ be the open ball of radius R in $\mathbb{R}^n$ centered at x, and let

(1.6)
$$u_{x,R} = \fint_{B_R} u(y)\,dy = \frac{1}{|B_R|}\int_{B_R(x)} u(y)\,dy\,.$$

Definition 1.2.1. Let μ be a constant, $0<\mu<1$. A function $u:\Omega \longrightarrow \mathbb{R}$ is *Hölder continuous* with exponent μ in Ω, if the quantity (Hölder seminorm)

(1.7)
$$[u]_\mu^\Omega = \sup_{\substack{x',x''\in\Omega\\ x'\neq x''}} \frac{|u(x')-u(x'')|}{|x'-x''|^\mu}$$

is finite. u is locally Hölder continuous in Ω, if u is Hölder continuous in every Ω', $\Omega' \subset\subset \Omega$. $C^{k,\mu}(\Omega)$ $(C^{k,\mu}_{loc}(\Omega))$ is the set of functions $u\in C^k(\Omega)$ $(C^k_{loc}(\Omega))$ whose k^{th} order derivatives are (locally) Hölder continuous with exponent μ in Ω $(k=0,1,2,\dots)$. The $C^{k,\mu}$–norm of u is

(1.8)
$$\|u\|_{C^{k,\mu}(\Omega)} = \|u\|_{C^k(\Omega)} + [D^k u]_\mu^\Omega = \|u\|_{C^k(\Omega)} + \sup_{|\alpha|=k}[D^\alpha u]_\mu^\Omega\,.$$

The following "Dirichlet growth theorem" by Morrey (see [MO 2]) guarantees Hölder continuity of certain Sobolev functions:

Theorem 1.2.2. *Let* $u\in W^{1,p}(\Omega)$, $1\le p<\infty$. *Suppose that for some constants* μ, M, $0<\mu\le 1$, $M>0$,

(1.9)
$$\int_{B_R} |Du|^p\,dy \le M^p R^{n-p+p\mu}$$

for all balls B_R *in* Ω. *Then* $u\in C^{0,\mu}_{loc}(\Omega)$, *and in each ball* B_R *with* $B_{3R}\subset\Omega$, *the estimate*

(1.10)
$$\operatorname*{osc}_{B_R} u \le C M R^\mu$$

holds with a constant C *which depends only on* n, p, μ.

Proof: By approximation, we may assume that $u\in C^1_{loc}(\Omega)$. Let $B_{3R}(x_0)\subset\Omega$ and let $x'\in B_R(x_0)$. Then $B_R(x_0)\subset B_{2R}(x')\subset\Omega$, and

$$|u(x')-u(x_0)| \le |u(x')-u_{x_0,R}| + |u(x_0)-u_{x_0,R}|$$
$$\le 2^n \fint_{B_{2R}(x')} |u(y)-u(x')|\,dy + \fint_{B_R(x_0)} |u(y)-u(x_0)|\,dy\,.$$

To estimate an integral of the form

$$\int_{B_R(x)} |u(y)-u(x)|\, dy,$$

note that

$$|u(y)-u(x)| \le \int_0^1 |Du(x+t(y-x))|\, dt\, |y-x|. \tag{1.11}$$

By integrating with respect to y over $B_R(x)$,

$$\begin{aligned}
\int_{B_R(x)} |u(y)-u(x)|\, dy &\le \int_0^1 \int_{B_R(x)} |Du(x+t(y-x))|\, |y-x|\, dy\, dt \\
&= \int_0^1 t^{-n-1} \int_{B_{tR}(x)} |Du(\xi)|\, |\xi-x|\, d\xi\, dt \\
&\le \int_0^1 t^{-n-1} \Big[\int_{B_{tR}} |Du|^p d\xi\Big]^{1/p} \Big[\int_{B_{tR}} |\xi-x|^{p/(p-1)} d\xi\Big]^{(p-1)/p} dt \\
&\le CM \int_0^1 t^{-n-1} (tR)^{(n-p+p\mu)/p} (tR)^{(n+p/(p-1))(p-1)/p} dt \\
&= CMR^{n+\mu} \int_0^1 t^{\mu-1} dt \\
&\le CMR^{n+\mu},
\end{aligned}$$

incorporating the assumed "Dirichlet growth." The theorem follows. □

1.3. Poincaré inequalities

Lemma 1.3.1. *Assume that* $u \in C^1(B_R)$, $B_R = B_R(x_0)$. *Then the inequality*

$$\int_{B_R} |u(y)-u(x)|^p dy \le CR^{n+p-1} \int_{B_R} |Du(y)|^p |y-x|^{1-n} dy \tag{1.12}$$

holds for all $x \in B_R$ *with a constant* C *which depends only on* n, p.

Proof: Recall (1.11), which implies, by Hölder's inequality, that

$$|u(y)-u(x)|^p \le \int_0^1 |Du(x+t(y-x))|^p dt\, |y-x|^p.$$

By integrating with respect to y over $\partial B_r(x) \cap B_R$, $0 < r < 2R$,

$$\int_{\partial B_r(x)\cap B_R} |u(y)-u(x)|^p \, d\sigma_y \le r^p \int_0^1 \int_{\partial B_r(x)\cap B_R} |Du(x+t(y-x))|^p \, d\sigma_y \, dt$$

$$= r^p \int_0^1 t^{1-n} \int_{\partial B_{tr}(x)\cap B_R} |Du(\xi)|^p \, d\sigma_\xi \, dt$$

$$= r^{n+p-2} \int_0^r \int_{\partial B_\tau(x)\cap B_R} |Du(\xi)|^p \, |\xi - x|^{1-n} \, d\sigma_\xi \, d\tau$$

$$\le r^{n+p-2} \int_{B_R} |Du(y)|^p \, |y-x|^{1-n} \, dy .$$

An integration with respect to r from 0 to $2R$ yields the stated inequality. □

<u>Theorem 1.3.2.</u> *There exists a constant* $C = C(n,p)$ *such that the Poincaré inequality*

$$\int_{B_R(x_0)} |u(x) - u_{x_0,R}|^p \, dx \le C R^p \int_{B_R(x_0)} |Du(x)|^p \, dx \tag{1.13}$$

holds for all $u \in W^{1,p}(B_R(x_0))$.

<u>Proof:</u> We may assume that $u \in C^1(B_R(x_0))$ and estimate by Lemma 1.3.1:

$$\int_{B_R(x_0)} |u(x) - u_{x_0,R}|^p \, dx = \int_{B_R} \left| ⨍_{B_R} (u(y)-u(x)) \, dy \right|^p dx$$

$$\le \int_{B_R} ⨍_{B_R} |u(y)-u(x)|^p \, dy \, dx$$

$$\le C R^{p-1} \int_{B_R} \int_{B_R} |Du(y)|^p \, |y-x|^{1-n} \, dy \, dx$$

$$\le C R^p \int_{B_R} |Du|^p \, dy$$

as required. □

<u>Theorem 1.3.3.</u> *The Poincaré inequality*

$$\int_{B_R} |u(x)|^p \, dx \le C R^p \int_{B_R} |Du(x)|^p \, dx \tag{1.14}$$

holds for all $u \in W_0^{1,p}(B_R)$ *with a constant* $C = C(n,p)$.

<u>Proof:</u> Let $u \in C_0^\infty(B_R)$ and choose x such that $u(x) = 0$. Lemma 1.3.1 then yields

$$|B_R| \int_{B_R} |u(y)|^p \, dy \le C R^{n+p-1} \int_{B_R} \int_{B_R} |Du(y)|^p \, |y-x|^{1-n} \, dy \, dx$$

$$\le C R^{n+p} \int_{B_R} |Du(y)|^p \, dy ,$$

from which the stated inequality follows. □

1.4. Integral characterizations of the Hölder spaces

The Poincaré inequality (1.13) hints towards a sharper criterion for Hölder continuity than the Dirichlet growth, namely that the oscillation integral growth

$$\int_{B_R(x)} |u(y) - u_{x,R}|^p \, dy \le K^p R^{n+p\mu}$$

be sufficient. μ–Hölder continuous functions obviously satisfy this condition. The following theorem by Campanato [CA 1] and Meyers [ME] provides therefore a precise integral characterization of the Hölder spaces:

Theorem 1.4.1. *Let* $u \in L^p(\Omega)$, $1 \le p < \infty$. *Assume that*

$$\int_{B_R(x)} |u(y) - u_{x,R}|^p \, dy \le K^p R^{n+p\mu} \tag{1.15}$$

for all balls $B_R(x)$ *in* Ω *with constants* μ, K, $0 < \mu \le 1$, $K > 0$.
Then $u \in C^{0,\mu}_{loc}(\Omega)$, *and there is a constant* $C = C(n, p, \mu)$ *such that the estimate*

$$\operatorname{osc}_{B_R} u \le C K R^{\mu} \tag{1.16}$$

holds for all balls B_R *with* $B_{3R} \subset \Omega$.

Proof: By the Lebesgue differentiation theorem,

$$u(x) = \lim_{R \to 0} u_{x,R}$$

for almost all $x \in \Omega$, and the functions $x \longmapsto u_{x,R}$ are continuous. Continuity of u would therefore follow from the claim that $\{u_{x,R}\}_{R>0}$ converges uniformly in $\Omega' \subset\subset \Omega$.
In order to prove this, let $R < R_0 < \operatorname{dist}(\Omega', \partial\Omega)$ and estimate

$$|u_{x,R} - u_{x,R_0}| \le |u(y) - u_{x,R}| + |u(y) - u_{x,R_0}| .$$

Raised to the pth power and integrated with respect to y over $B_R(x)$, this inequality yields

$$|B_R|\,|u_{x,R}-u_{x,R_0}|^p \le 2^{p-1}\left\{\int_{B_R(x)}|u(y)-u_{x,R}|^p dy + \int_{B_{R_0}(x)}|u(y)-u_{x,R_0}|^p dy\right\}$$

$$\le 2^p K^p R_0^{n+p\mu},$$

and therefore

$$|u_{x,R}-u_{x,R_0}| \le C K R_0^{\frac{n}{p}+\mu} R^{-\frac{n}{p}}$$

Let $R_k = R/2^k$, $(k = 0,1,2,\dots)$. Then

$$|u_{x,R_{k+1}}-u_{x,R_k}| \le C K R^\mu 2^{-k\mu},$$

and, for $k > k_0$, by summation,

$$|u_{x,R_k}-u_{x,R_{k_0}}| \le C K R^\mu \sum_{m=k_0}^{k-1} 2^{-m\mu}$$

(1.17)

$$\le C K R_{k_0}^\mu .$$

The sequence $\{u_{x,R_k}\}_{k=1}^\infty$ is therefore a uniform Cauchy sequence in $\Omega' \subset\subset \Omega$. The limit

$$\bar{u}(x) = \lim_{k\to\infty} u_{x,R_k}$$

exists for all $x \in \Omega$ and is a continuous function which equals $u(x)$ almost everywhere. This also implies that $\bar{u}$ does not depend on the choice of R, and we can take $\bar{u}$ as the precise representative of u. Furthermore, by putting $k_0 = 0$ in (1.17) and letting $k \to \infty$,

$$|u(x)-u_{x,R}| \le C K R^\mu. \tag{1.18}$$

$\{u_{x,R}\}_{R>0}$ converges therefore uniformly to u in $\Omega' \subset\subset \Omega$.

To show Hölder continuity of u, let $B_{3R}(x_0) \subset \Omega$ and let $x' \in B_R(x_0)$, so that $B_R(x_0) \subset R_{2R}(x') \subset \Omega$. Then estimate

$$|u(x')-u(x_0)| \le |u(x')-u_{x',2R}| + |u(x_0)-u_{x_0,R}| + |u_{x',2R}-u_{x_0,R}|$$

(1.19)

$$\le C K R^\mu + |u_{x',2R}-u_{x_0,R}|$$

by (1.18). Now, as before,

$$|B_R|\,|u_{x',2R}-u_{x_0,2R}|^p \le 2^{p-1}\left\{\int_{B_{2R}(x')}|u(y)-u_{x',2R}|^p\,dy+\int_{B_R(x_0)}|u(y)-u_{x_0,R}|^p\,dy\right\}$$
$$\le 2^p K^p R^{n+p\mu}.$$

Hence

$$|u_{x',2R}-u_{x_0,R}| \le C K R^{\mu},$$

which, combined with (1.19) yields

$$|u(x')-u(x'')| \le C K R^{\mu}$$

whenever $x', x'' \in B_R(x_0)$ as required. □

Theorem 1.4.2. *The statement of Theorem* 1.4.1 *remains true, if for any ball* $B_R(x)$ *in* Ω, *there is a constant* $c_{x,R}$ *such that*

$$\int_{B_R(x)}|u(y)-c_{x,R}|^p\,dy \le K^p R^{n+p\mu}. \tag{1.20}$$

Proof: We estimate

$$\int_{B_R(x)}|u(y)-u_{x,R}|^p\,dy \le 2^{p-1}\left\{\int_{B_R(x)}|u(y)-c_{x,R}|^p\,dy+\int_{B_R(x)}|u_{x,R}-c_{x,R}|^p\,dy\right\}$$
$$\le C\left\{K^pR^{n+p\mu}+R^{n-np}\left|\int_{B_R}(u(y)-c_{x,R})\,dy\right|^p\right\}$$
$$\le C\left\{K^pR^{n+p\mu}+\int_{B_R}|u(y)-c_{x,R}|^p\,dy\right\}$$
$$\le CK^pR^{n+p\mu},$$

incorporating Hölder's inequality and the assumption (1.20). The statement follows from Theorem 1.4.1. □

A further extension of Theorem 1.4.1 can be found in Campanato [CA 2]. We however content ourselves with a particularly simple case. The following De Giorgi lemma is useful:

Lemma 1.4.3. *Let* $P(x)=\sum c_\alpha x^\alpha$, $x^\alpha = x_1^{\alpha_1}\cdot\ldots\cdot x_n^{\alpha_n}$ *be a polynomial of degree* $\le k$. *Then*

$$|c_\alpha|^p R^{n+p|\alpha|} \le C\int_{B_R(x_0)}|P(x-x_0)|^p\,dx \tag{1.21}$$

for $|\alpha|\le k$ *with a constant* $C=C(n,k,p)$.

Proof: Let $\mathscr{P}_n^k$ consist of all polynomials $P(x) = \sum c_\alpha x^\alpha$ of degree $\le k$ with $\sum |c_\alpha|^2 = 1$. Consider the variational problem

$$d = d(n,k,p) = \inf_{P \in \mathscr{P}_n^k} \int_{B_1(0)} |P(x)|^p dx.$$

Let $P^\nu(x) = \sum c_\alpha^\nu x^\alpha$, $\nu = 1,2,\ldots$, be a minimizing sequence, i.e.,

$$d \le \int_{B_1(0)} |P^\nu(x)|^p dx \le d + \frac{1}{\nu}.$$

Then $\left\{ \left[c_\alpha^\nu\right]_{|\alpha| \le k} \right\}_{\nu=1}^\infty$ possesses a convergent subsequence $\left[c_\alpha^{\nu_\mu}\right]_{|\alpha|\le k} \longrightarrow \left[c_\alpha^0\right]_{|\alpha|\le k}$ as $\mu \longrightarrow \infty$. Hence $P^{\nu_\mu}(x) \longrightarrow P^0(x) = \sum c_\alpha^0 x^\alpha$ uniformly on $B_1(0)$, and $P^0 \in \mathscr{P}_n^k$. Therefore

$$d = \min_{P \in \mathscr{P}_n^k} \int_{B_1(0)} |P(x)|^p dx = \int_{B_1(0)} |P^0(x)|^p dx > 0.$$

The inequality

$$d\Big[\sum_{|\alpha|\le k} |c_\alpha|^2 \Big]^{p/2} \le \int_{B_1(0)} |P(x)|^p dx$$

follows for arbitrary $P(x) = \sum c_\alpha x^\alpha$. Finally

$$\begin{aligned} \int_{B_R(x_0)} |P(x-x_0)|^p dx &= R^n \int_{B_1(0)} |P(Ry)|^p dy \\ &\ge d\Big[\sum_{|\alpha|\le k} |c_\alpha R^{|\alpha|}|^2 \Big]^{p/2} R^n \\ &\ge d\, |c_\alpha|\, R^{n+p|\alpha|} \end{aligned}$$

for $|\alpha| \le k$ as required. □

Proposition 1.4.4. *Let $u \in L^p(\Omega)$, $1 \le p < \infty$, and let $0 < \mu \le 1$, $K > 0$. Suppose that for each ball $B_R(x)$ in Ω, there is a polynomial $P_{x,R}(y) = \sum c_{\alpha,x,R} y^\alpha$ of degree $\le k$ with*

(1.22) $$\int_{B_R(x)} |u(y) - P_{x,R}(y-x)|^p dy \le K^p R^{n+p\mu}.$$

Assume further that $x \longmapsto c_{0,x,R}$ is continuous for given R and that, for almost all x,

(1.23) $$|c_{\alpha,x,R}| \le c_x < \infty$$

for $0 < R \le \epsilon_x$ for some $\epsilon_x > 0$ and $1 \le |\alpha| \le k$.

Then $u \in C^0_{loc}(\Omega)$, *and the estimate*

$$|u(x)-c_{0,x,R}| \le C K R^{\mu} \tag{1.24}$$

holds whenever $B_R(x) \subset \Omega$ *with* $C = C(n,p,\mu)$.

<u>**Proof:**</u> Let $B_{R_0} \subset \Omega$ and let $R < R_0$. Then estimate

$$\begin{aligned}
&\int_{B_R(x)} |P_{x,R}(y-x)-P_{x,R_0}(y-x)|^p \, dy \\
&\quad \le 2^{p-1}\left\{\int_{B_R(x)} |u(y)-P_{x,R}(y-x)|^p \, dy + \int_{B_{R_0}(x)} |u(y)-P_{x,R_0}(y-x)|^p \, dy\right\} \\
&\quad \le 2^p K^p R_0^{n+p\mu}.
\end{aligned}$$

Invoking Lemma 1.4.3 and writing $c_{x,R} = c_{0,x,R}$, it follows that

$$|c_{x,R}-c_{x,R_0}| \le C K R_0^{\frac{n}{p}+\mu} R^{-\frac{n}{p}}.$$

Let $R_k = R/2^k$. In following the proof of Theorem 1.4.1, one shows that

$$c(x) = c_R(x) = \lim_{k\to\infty} c_{x,R_k}$$

exists as a continuous function and

$$|c(x)-c_{x,R}| \le C K R^{\mu}. \tag{1.25}$$

To show that $c(x) = u(x)$ a. e., estimate

$$\begin{aligned}
&|B_R|\,|u(x)-c(x)|^p \\
&\quad \le 2^{2p-2}\left\{\int_{B_R(x)} |u(y)-u(x)|^p \, dy + \int_{B_R(x)} |u(y)-P_{x,R}(y-x)|^p \, dy\right. \\
&\qquad \left. + \int_{B_R(x)} |P_{x,R}(y-x)-c_{x,R}|^p \, dy + |B_R|\,|c(x)-c_{x,R}|^p\right\} \\
&\quad \le C\left\{\int_{B_R(x)} |u(y)-u(x)|^p \, dy + \int_{B_R(x)} |P_{x,R}(y-x)-c_{x,R}|^p \, dy + K R^{n+p\mu}\right\}
\end{aligned}$$

by the assumption (1.22) and by (1.25). Now

$$\fint_{B_R(x)} |u(y)-u(x)|^p\,dy \longrightarrow 0 \quad \text{as } R \longrightarrow 0$$

for a.e. x and, by assumption (1.23),

$$\fint_{B_R(x)} |P_{x,R}(y-x)-c_{x,R}|^p\,dy \le C \max_{|\alpha|\le k} |c_{\alpha,x,R}| \sum_{\ell=1}^{k} R^{\ell}$$

$$\longrightarrow 0 \quad \text{as} \quad R \longrightarrow 0 .$$

Therefore $c(x) = u(x)$ a.e. and the stated inequality (1.24) holds for all x. □

1.5. A Sobolev inequality

Lemma 1.5.1. *Let* $u \in W^{k,p}(\Omega)$, $k = 1,2,\dots$, $1 \le p < \infty$. *Then for each ball* $B_R(x)$ *in* Ω, *there is a polynomial* $P_{x,R}(y) = P^{k-1}_{u,x,R}(y) = \sum c_{\alpha,x,R}\, y^{\alpha}$ *of degree* $\le k-1$ *with*

$$\int_{B_R(x)} |u(y)-P_{x,R}(y-x)|^p\,dy \le C R^{kp} \int_{B_R(x)} |D^k u|^p\,dy . \tag{1.26}$$

Furthermore, $x \longmapsto c_{\alpha,x,R}$ *is continuous for fixed* R *and, for almost all* x,

$$|c_{\alpha,x,R}| \le c_x < \infty \tag{1.27}$$

for all $R \le \operatorname{dist}(x, \partial\Omega)$. *Moreover the inequality*

$$|c_{\alpha,x,R}| \le \frac{C}{R^n} \|u\|_{W^{k-1,p}(B_R)} \tag{1.28}$$

holds for all $|\alpha| \le k$. *The constant* C *depends only on* n, k, p.

Proof: For $k = 1$, the statement follows from Poincaré's inequality (1.13). Assume by induction that (1.26) is true for a $k \ge 1$. Let $u \in W^{k+1,p}(\Omega)$ and set

$$\tilde{u}(y) = u(y) - \sum_{|\alpha|=k} \frac{1}{\alpha!} (D^{\alpha}u)_{x,R}\, (y-x)^{\alpha} .$$

Then there is a polynomial $P^{k-1}_{\tilde{u}}(y)$ of degree $\le k-1$ such that

$$\int_{B_R(x)} |\tilde{u}(y)-P^{k-1}_{\tilde{u}}(y-x)|^p\,dy \le C R^{kp} \int_{B_R(x)} |D^k \tilde{u}|^p\,dy$$

$$= C R^{kp} \sum_{|\alpha|=k} \int_{B_R(x)} |D^{\alpha}u-(D^{\alpha}u)_{x,R}|^p\,dy$$

$$\le C R^{(k+1)p} \int_{B_R(x)} |D^{k+1}u|^p \, dy .$$

This by virtue of the induction assumption and by Poincaré's inequality (1.13). Let

$$P_u^k(y) = \sum_{|\alpha|=k} \frac{1}{\alpha!} (D^\alpha u)_{x,R}\, y^\alpha + P_{\tilde u}^{k-1}(y) .$$

Then

$$\int_{B_R(x)} |\tilde u(y) - P_{\tilde u}^{k-1}(y-x)|^p \, dy = \int_{B_R(x)} |u(y) - P_u^k(y-x)|^p \, dy .$$

The additional properties follow easily, in particular (1.27) by the Lebesgue differentiation theorem. This completes the induction. □

Theorem 1.5.2. *Let* $u \in W^{k,p}(\Omega)$, $k = 1,2,\dots$, $1 \le p < \infty$, *with* $k - \frac{n}{p} > 0$. *Then* $u \in C^0_{loc}(\Omega)$, *and the Sobolev inequality*

$$\sup_{B_R} |u| \le C(R) \|u\|_{W^{k,p}(B_{2R})} \tag{1.29}$$

holds for all B_R *with* $B_{2R} \subset \Omega$. *The constant* $C(R)$ *depends in addition to* R *also on* n, k, p.

Proof: Assume w.l.o.g. that $\mu = k - \frac{n}{p} \le 1$, and let $B_R(x) \subset \Omega$. By Lemma 1.5.1 there is a polynomial $P_{x,R}(y)$ of degree $\le k-1$ such that

$$\int_{B_R(x)} |u(y) - P_{x,R}(y-x)|^p \, dy \le C R^{kp} \int_{B_R(x)} |D^k u|^p \, dy .$$

$P_{x,R}(y)$ satisfies the technical assumptions of Proposition 1.4.4, from which $u \in C^0_{loc}(\Omega)$ and

$$|u(x) - c_{x,R}| \le C R^\mu \|D^k u\|_{L^p(B_R)} .$$

By (1.28) we obtain the bound

$$|u(x)| \le C(R) \|u\|_{W^{k,p}(B_R)} ,$$

and a simple covering argument yields the stated inequality (1.29). □

1.6. The Courant–Lebesgue Lemma

Let Ω be an open set in $\mathbb{R}^2$. The following lemma (see Courant [CO]) provides an estimate for the continuity modulus of homeomorphisms $u = (u^1, u^2)$ with finite Dirichlet integral:

Lemma 1.6.1. *Suppose that* $u:\Omega \longrightarrow u(\Omega)$ *is a homeomorphism of class* $W^{1,2}(\Omega,\mathbb{R}^2)$ *and let*

$$\int_\Omega |Du|^2 dx \le M. \tag{1.30}$$

Then, for any B_R *with* $B_{\sqrt{R}} \subset \Omega$, $0 < R < 1$,

$$\operatorname*{osc}_{B_R} u \le 2\sqrt{\frac{\pi M}{\log 1/R}}. \tag{1.31}$$

Proof: Let $B_{\sqrt{R}}(x_0) \subset \Omega$, $0 < R < 1$. By changing to polar coordinates (r,θ) about x_0, one computes

$$\int_R^{\sqrt{R}}\int_0^{2\pi}\left[|u_r|^2+\frac{1}{r^2}|u_\theta|^2\right] r\,d\theta dr = \int_{B_{\sqrt{R}}\setminus B_R} |Du|^2 dx \le M,$$

and therefore

$$\int_R^{\sqrt{R}}\frac{1}{r}\int_0^{2\pi}|u_\theta|^2 d\theta dr \le M.$$

Assume that

$$\int_0^{2\pi}|u_\theta(r,\theta)|^2 d\theta > \frac{2M}{\log 1/R}$$

for a.e. r, $R \le r \le \sqrt{R}$. Then

$$\int_R^{\sqrt{R}}\frac{1}{r}\int_0^{2\pi}|u_\theta(r,\theta)|^2 d\theta dr > \frac{2M}{\log 1/R}\int_R^{\sqrt{R}}\frac{1}{r}dr = M,$$

a contradiction. Hence there is an R^*, $R \le R^* \le \sqrt{R}$, such that $\theta \longmapsto u(R^*,\theta)$ is absolutely continuous on $[0,2\pi]$, and such that

$$\int_0^{2\pi}|u_\theta(R^*,\theta)|^2 d\theta \le \frac{2M}{\log 1/R}.$$

We conclude that for all θ_1, θ_2, $0 \le \theta_1 \le \theta_2 \le 2\pi$,

$$\begin{aligned}|u(R^*,\theta_2)-u(R^*,\theta_1)| &\le \int_{\theta_1}^{\theta_2}|u_\theta(R^*,\theta)|\,d\theta\\ &\le \sqrt{2\pi}\left[\int_{\theta_1}^{\theta_2}|u_\theta(R^*,\theta)|^2 d\theta\right]^{1/2}\\ &\le 2\sqrt{\frac{\pi M}{\log 1/R}},\end{aligned}$$

from which

$$\operatorname*{osc}_{\partial B_{R^*}} u \le 2\sqrt{\frac{\pi M}{\log 1/R}}.$$

Remembering that $R^* \ge R$, the statement of the lemma follows from the homeomorphic character of the mapping u. □

__Corollary 1.6.2.__ *Suppose that* $\Omega = B = B_1(0)$, *and let* $x_0 \in \bar{B}$, $0 < R < 1$. *Then there is a radius* R^* *with* $R \le R^* \le \sqrt{R}$ *such that*

$$\operatorname*{osc}_{\partial B_{R^*}(x_0)\cap B} u \le 2\sqrt{\frac{\pi M}{\log 1/R}}. \tag{1.32}$$

__Lemma 1.6.3.__ *Let* u *be a homeomorphism from the closed disc* $\bar{B}$ *onto itself of class* $W^{1,2}(B)$ *such that* $u(0) = 0$ *and*

$$\int_B |Du|^2 dx \le M. \tag{1.33}$$

Then

$$|u(x')-u(x'')| \le 4\sqrt{\frac{\pi M}{\log 1/R}} \tag{1.34}$$

for any $x', x' \in B$ *such that* $|x'-x''| \le R$, $0 < R < 1/4$.

__Proof:__ We may assume that

$$4\sqrt{\frac{\pi M}{\log 1/R}} < 2.$$

Two possibilities may occur when Corollary 1.6.2 is applied: If $B_{R^*}(x_0) \subset B$, then

$$\operatorname*{osc}_{B_R} u \le \operatorname*{osc}_{B_{R^*}} u \le \operatorname*{osc}_{\partial B_{R^*}} u \le 2\sqrt{\frac{\pi M}{\log 1/R}},$$

because u is a homeomorphism.

If $B_{R^*}(x_0)\cap \partial B \ne \emptyset$, then $0 \notin B_{R^*}(x_0)$ because $R^* \le \sqrt{R} \le 1/2$. Pick $\tilde{u} \in u(\partial B_{R^*}\cap B)\cap \partial B$ and let

$$\bar{B}(\tilde{u}) = \bar{B}_{\tilde{R}}(\tilde{u}), \qquad \tilde{R} = 2\sqrt{\frac{\pi M}{\log 1/R}}.$$

By Corollary 1.6.2,

$$u(\partial B_{R^*}\cap B) \subset \bar{B}(\tilde{u}).$$

We claim that

$$u(B_{R^*} \cap B) \subset \bar{B}(\tilde{u}).$$

Assume by contradiction, that there is a point $x^* \in B_{R^*} \cap B$ such that $u^* = u(x^*) \notin \bar{B}(\tilde{u})$. Since $\tilde{R} < 1$, we have $0 \notin \bar{B}(\tilde{u})$. Hence there exists a curve $\gamma \subset \bar{B} \setminus \bar{B}(\tilde{u})$ joining 0 and u^*. Now $u^{-1}(\gamma)$ joins 0 and x^*, and, because $0 \notin B_{R^*} \cap B$, there exists a $\tilde{x} \in u^{-1}(\gamma) \cap \partial B_{R^*} \cap B$. On the other hand, $u(\tilde{x}) \notin \bar{B}(\tilde{u}) \supset u(\partial B_{R^*} \cap B)$, a contradiction.

Suppose now that $x_0, x_1 \in B$, $|x_0 - x_1| \leq R$, $0 < R < 1/4$. Then

$$|u(x_0) - u(x_1)| \leq |u(x_0) - \tilde{u}| + |u(x_1) - \tilde{u}|$$

$$\leq 4 \sqrt{\frac{\pi M}{\log 1/R}}$$

also in the case $B_{R^*}(x_0) \cap \partial B \neq \emptyset$ as required. □

Chapter 2. REGULARITY FOR LINEAR ELLIPTIC EQUATIONS AND QUASILINEAR SYSTEMS

In this chapter we shall describe the regularity theory for linear elliptic equations of divergence structure and quasilinear systems of diagonal form only to the extent that we need it. The theory was developed by Morrey [MO 1] and Campanato [CA 3, 4] and is usually referred to as the Campanato technique.

2.1. Linear homogeneous equations with constant coefficients

Consider the elliptic equation

$$-D_\beta(a^{\alpha\beta}D_\alpha u) = 0 \tag{2.1}$$

with constant coefficient matrix $[a^{\alpha\beta}]^n_{\alpha,\beta=1}$, which is assumed to be real, symmetric and positive definite. Let λ, Λ be positive constants such that

$$\lambda\,|\xi|^2 \le a^{\alpha\beta}\xi_\alpha\xi_\beta \le \Lambda\,|\xi|^2 \tag{2.2}$$

for all $\xi = (\xi_1,\dots,\xi_n) \in \mathbb{R}^n$.

A first step towards regularity of $W^{1,2}$–solutions is the Caccioppoli inequality:

Lemma 2.1.1. *Let* $u \in W^{1,2}(\Omega)$ *be a weak solution of the equation* (2.1), *i. e.*

$$\int_\Omega a^{\alpha\beta}D_\alpha u\, D_\beta\eta\, dx = 0 \tag{2.3}$$

for all $\eta \in C_0^\infty(\Omega)$, *resp.* $\eta \in W_0^{1,2}(\Omega)$.

Then there is a constant $C = C(n,\lambda,\Lambda)$ *such that the inequality*

$$\int_{B_R} |Du|^2 dx \le \frac{C}{R^2}\int_{B_{2R}} |u|^2 dx \tag{2.4}$$

holds for all balls B_R *with* $B_{2R} \subset \Omega$.

Proof: Let $\zeta \in C_0^\infty(B_{2R})$ be a function with the properties $0 \le \zeta \le 1$, $\zeta \equiv 1$ on B_R, $|D\zeta| \le c(n)/R$. Take $\eta = u\,\zeta^2$ as a test function in (2.3) to obtain

$$\int_{B_{2R}} a^{\alpha\beta}D_\alpha u\, D_\beta u\, \zeta^2 dx + 2\int_{B_{2R}} a^{\alpha\beta}D_\alpha u\, u\,\zeta\, D_\beta\zeta\, dx = 0\,.$$

Hence

$$\lambda\int_{B_{2R}}|Du|^2\zeta^2\,dx \le \frac{C(n,\Lambda)}{R}\int_{B_{2R}}|Du|\,|u|\,\zeta\,dx$$
$$\le C(n,\Lambda)\,\epsilon\int_{B_{2R}}|Du|^2\zeta^2\,dx + \frac{1}{\epsilon R^2}\int_{B_{2R}}|u|^2\,dx.$$

Choose ϵ small to obtain the stated inequality. □

The difference quotient method is used to obtain higher regularity for u:

<u>**Proposition 2.1.2.**</u> *Let* $u \in W^{1,2}(\Omega)$ *be a weak solution of* (2.1). *Then* $u \in W^{2,2}_{loc}(\Omega)$, *and* $D_i u$ *solves* (2.1) *for all* $i = 1,\ldots,n$.

<u>**Proof:**</u> For $i = 1,\ldots,n$ let $e_i = (\delta_{ij})_{j=1}^n$ be the ith standard unit vector in $\mathbb{R}^n$. For $\eta \in C_0^\infty(\Omega)$ let $h \in \mathbb{R}$, $|h| < \operatorname{dist}(\operatorname{supp}\eta, \partial\Omega)$. Then

$$\int_\Omega a^{\alpha\beta} D_\alpha u(x+h\,e_i)\,D_\beta\eta(x)\,dx = \int_\Omega a^{\alpha\beta} D_\alpha u(x)\,D_\beta(\eta(x-h\,e_i))\,dx$$
$$= 0.$$

Hence

$$\int_\Omega a^{\alpha\beta} D_\alpha(\Delta_i^h u)\,D_\beta\eta\,dx = 0, \tag{2.5}$$

where

$$\Delta_i^h u(x) = \frac{u(x+h\,e_i)-u(x)}{h}$$

is the difference quotient of u in the direction e_i.

Caccioppoli's inequality (2.4) is therefore satisfied by $\Delta_i^h u$, and whence

$$\int_{B_R}|D(\Delta_i^h u)|^2\,dx \le \frac{C}{R^2}\int_{B_{2R}}|\Delta_i^h u|^2\,dx$$
$$\le \frac{C}{R^2}\int_{B_{3R}}|Du|^2\,dx,$$

if $B_{3R} \subset \Omega$, $|h| < R$. The latter inequality follows by estimating

$$|\Delta_i^h u|^2 = \left|\frac{1}{h}\int_0^h D_i u(x+t\,e_i)\,dt\right|^2$$
$$\le \frac{1}{h}\int_0^h |D_i u(x+t\,e_i)|^2\,dt,$$

from which

$$\int_{B_{2R}} |\Delta_i^h u|^2 dx \le \frac{1}{h}\int_0^h \int_{B_{2R}} |D_i u(x+te_i)|^2 dx\,dt$$
$$\le \int_{B_{3R}} |Du|^2 dx .$$

The inequality

$$\int_{B_R} |\Delta_i^h (Du)|^2 dx \le \frac{C}{R^2}\int_{B_{3R}} |Du|^2 dx$$

implies the existence of the weak derivatives $D_{ij}u$ for $i,j = 1,\dots,n$ together with the estimate

(2.6) $$\int_{B_R} |D^2 u|^2 dx \le \frac{C}{R^2}\int_{B_{3R}} |Du|^2 dx .$$

This is true because of the weak compactness of bounded sets in L^2. The derivatives $D_i u$ solve therefore the equation (2.3) for $i = 1,\dots,n$ as required. □

The repeated application of Proposition 2.1.2 together with a simple convering argument and the Sobolev embedding theorem, Theorem 1.5.2, gives

Theorem 2.1.3. *Let* $u \in W^{1,2}(\Omega)$ *be a weak solution of* (2.1). *Then* $u \in C^\infty_{loc}(\Omega)$ *and for all balls* B_R *with* $B_{2R} \subset \Omega$, *and all* k, $k = 1,2,\dots,$

(2.7) $$\|u\|_{W^{k,2}(B_R)} \le C(n,k,\lambda,\Lambda,R)\|u\|_{L^2(B_{2R})} .$$

Theorem 2.1.4. *Let* $u \in W^{1,2}(B)$, $B = B_{R_0}(x_0)$, *be a solution of* (2.1). *Then for all* R, $0 < R \le R_0$,

(2.8) $$\int_{B_R(x_0)} |u|^2 dx \le C\left[\frac{R}{R_0}\right]^n \int_{B_{R_0}(x_0)} |u|^2 dx ,$$

(2.9) $$\int_{B_R(x_0)} |u-u_{x_0,R}|^2 dx \le C\left[\frac{R}{R_0}\right]^{n+2} \int_{B_{R_0}(x_0)} |u-u_{x_0,R_0}|^2 dx$$

with a constant $C = C(n,\lambda,\Lambda)$.

Proof: We employ the regularity theorem, Theorem 2.1.3, the Sobolev inequality (1.29) for $k = \left[\frac{n}{2}\right] + 1$ and the estimate (2.7) to obtain, for $R \le R_0/4$,

$$\int_{B_R(x_0)} |u|^2 dx \le C R^n \sup_{B_{R_0/4}} |u|^2$$

$$\le C(R_0) R^n \|u\|^2_{W^{k,2}(B_{R_0/2})}$$

$$\le C(R_0) R^n \int_{B_{R_0}} |u|^2 dx .$$

Such an inequality is easily seen to hold also for $R_0/4 \le R \le R_0$. The stated R_0-dependence is obtained by a simple scaling argument. This proves inequality (2.8).

The inequality (2.9) is true for $R \le R_0/2$, because Du, $u - u_{x_0,R}$ solve (2.1), and hence

$$\int_{B_R(x_0)} |Du|^2 dx \le C \left[\frac{R}{R_0}\right]^n \int_{B_{R_0/2}(x_0)} |Du|^2 dx \tag{2.10}$$

$$\le C \left[\frac{R}{R_0}\right]^n \frac{1}{R_0^2} \int_{B_{R_0}(x_0)} |u - u_{x_0,R_0}|^2 dx ,$$

incorporating Caccioppoli's inequality (2.4). The Poincaré inequality (1.13), applied to the left hand side of (2.10), yields the stated inequality (2.9), which is easily seen to hold also in the case $R_0/2 \le R \le R_0$. □

2.2. Linear equations with constant coefficients

Consider a solution $u \in W^{1,2}(\Omega)$ of an elliptic equation with constant coefficients of the form

$$-D_\beta(a^{\alpha\beta} D_\alpha u) = -D_\alpha f^\alpha + g . \tag{2.11}$$

Theorem 2.2.1. *Suppose that* $f \in C^\mu(\Omega, \mathbb{R}^n)$ *for some* μ, $0 < \mu < \infty$, $g \in L^\infty(\Omega)$, *and let* $u \in W^{1,2}(\Omega)$ *solve* (2.11). *Then* $u \in C^{1,\mu}_{loc}(\Omega)$, *and an estimate of the form*

$$\operatorname{osc}_{B_R} Du \le C R^\mu \left\{ \|Du\|_{L^2(\Omega)} + [f]^\Omega_\mu + \|g\|_{L^\infty(\Omega)} \right\} \tag{2.12}$$

holds for all balls $B_R(x_0)$ *with* $B_{3R} \subset \Omega$ *and with a constant* C *which depends only on* n, μ, λ, Λ, $\operatorname{dist}(x_0, \partial\Omega)$.

Proof: Let $B_{R_0} = B_{R_0}(x_0) \subset \Omega$ and let $v \in W^{1,2}(B_{R_0})$ be the (weak) solution of the Dirichlet problem

$$-D_\beta(a^{\alpha\beta} D_\alpha v) = 0 \quad \text{in } B_{R_0}, \qquad v - u \in W^{1,2}_0(B_{R_0}) . \tag{2.13}$$

Theorem 2.1.4, applied to Dv, gives for all R, $0 < R \le R_0$:

$$\int_{B_R(x_0)} |Dv-(Dv)_{x_0,R}|^2 dx \le C\left[\frac{R}{R_0}\right]^{n+2} \int_{B_{R_0}(x_0)} |Dv-(Dv)_{x_0,R_0}|^2 dx. \tag{2.14}$$

On the other hand, it follows from (2.11) and (2.13) that the integral relation

$$\int_{B_{R_0}} a^{\alpha\beta} D_\alpha(u-v)\, D_\beta \eta\, dx = \int_{B_{R_0}} (f^\alpha - f^\alpha(x_0))\, D_\alpha \eta\, dx + \int_{B_{R_0}} g\,\eta\, dx$$

holds for all $\eta \in W_0^{1,2}(B_{R_0})$. By letting $\eta = u-v$ and by employing the ellipticity and the Poincaré inequality (1.14), we obtain

$$\begin{aligned}\lambda\int_{B_{R_0}} |D(u-v)|^2 dx &\le \epsilon\int_{B_{R_0}} |D(u-v)|^2 dx + \frac{C}{\epsilon}\int_{B_{R_0}} |f(x)-f(x_0)|^2 dx \\ &\quad + \frac{\epsilon}{R_0^2}\int_{B_{R_0}} |u-v|^2 dx + \frac{R_0^2}{\epsilon}\int_{B_{R_0}} |g|^2 dx \\ &\le \epsilon C\int_{B_{R_0}} |D(u-v)|^2 dx + \frac{C}{\epsilon} R_0^{n+2\mu}.\end{aligned}$$

By choosing ϵ to be small,

$$\int_{B_{R_0}(x_0)} |D(u-v)|^2 dx \le C R_0^{n+2\mu}. \tag{2.15}$$

Combined with (2.14), and incorporating the Hölder inequality, this estimate gives

$$\begin{aligned}\int_{B_R(x_0)} |Du-(Du)_{x_0,R}|^2 dx &\le 2^{2p-2}\left\{\int_{B_R} |Dv-(Dv)_{x_0,R}|^2 dx + 2\int_{B_R} |D(u-v)|^2 dx\right\} \\ &\le C\left\{\left[\frac{R}{R_0}\right]^{n+2}\int_{B_{R_0}(x_0)} |Du-(Du)_{x_0,R_0}|^2 dx + R_0^{n+2\mu}\right\}\end{aligned}$$

for all R, R_0, $0 < R \le R_0 \le \operatorname{dist}(x_0, \partial\Omega)$. This implies by the following Lemma 2.2.2 that

$$\begin{aligned}\int_{B_R(x_0)} |Du-(Du)_{x_0,R}|^2\, dx &\le C\left\{\left[\frac{R}{R_0}\right]^{n+2\mu}\int_{B_{R_0}(x_0)} |Du-(Du)_{x_0,R_0}|^2 dx + R^{n+2\mu}\right\} \\ &\le C R^{n+2\mu}\left\{\frac{1}{R_0^{n/2+\mu}}\|Du\|_{L^2(\Omega)} + [f]_\mu^\Omega + \|g\|_{L^\infty(\Omega)}\right\}^2,\end{aligned}$$

as a careful analysis of C shows. Theorem 1.4.1 implies the stated estimate and thus the regularity $u \in C^{1,\mu}_{loc}(\Omega)$ as required. □

<u>Lemma 2.2.2.</u> *Let* $\phi(R)$ *be a nonnegative, nondecreasing function on* $0 < R \le R_1$. *Suppose that*

$$\phi(R) \le A\left[\left[\frac{R}{R_0}\right]^{\alpha} + \epsilon\right]\phi(R_0) + B R_0^{\beta} \tag{2.16}$$

for all R, R_0, $0 < R \le R_0 \le R_1$, *with* $A, B, \epsilon, \alpha, \beta$ *being nonnegative constants,* $A \ge 1$, $\beta < \alpha$. *Then there exists a constant* $\epsilon_0 = \epsilon_0(A, \alpha, \beta)$ *such that*

$$\phi(R) \le C\left[\left[\frac{R}{R_0}\right]^{\beta}\phi(R_0) + B R^{\beta}\right] \tag{2.17}$$

for all $0 < R \le R_0 \le R_1$ *with* $C = C(A, \alpha, \beta)$, *if* $\epsilon \le \epsilon_0$ *in* (2.16).

<u>Proof:</u> Let τ, $0 < \tau < 1$, be the solution of $2A\tau^{\alpha} = \tau^{\gamma}$, where $\gamma = (\alpha+\beta)/2$. Put $\epsilon_0 = \tau^{\alpha}$. Then we have, for $\epsilon \le \epsilon_0$,

$$\begin{aligned}\phi(\tau R_0) &\le A(\tau^{\alpha} + \epsilon)\,\phi(R_0) + B R_0^{\beta} \\ &\le \tau^{\gamma}\phi(R_0) + B R_0^{\beta}.\end{aligned}$$

This inequality is now iterated to yield, for $m = 1, 2, \ldots$,

$$\begin{aligned}\phi(\tau^m R_0) &\le \tau^{\gamma}\phi(\tau^{m-1}R_0) + B(\tau^{m-1}R_0)^{\beta} \\ &\le \tau^{m\gamma}\phi(R_0) + B R_0^{\beta}\sum_{\ell=0}^{m-1}\tau^{(m-1-\ell)\beta}\tau^{\ell\gamma} \\ &\le \tau^{m\gamma}\phi(R_0) + B(\tau^{m-1}R_0)^{\beta}\sum_{\ell=0}^{m-1}\tau^{\ell(\gamma-\beta)} \\ &\le C\tau^{m\gamma}(\phi(R_0) + B R_0^{\beta}).\end{aligned}$$

Now let $R \le R_0$, and choose $m = 0, 1, 2, \ldots$, so that $\tau^{m+1}R_0 < R \le \tau^m R_0$, to obtain the stated inequality (2.17). □

Regarding the monotonicity of

$$\phi(x_0, R) = \int_{B_R(x_0)} |Du - (Du)_R|^2\,dx, \tag{2.18}$$

the following lemma is useful:

Lemma 2.2.3. *Let* $u \in L^2(B_R)$. *Then for all* $c \in \mathbb{R}$,

(2.19)
$$\int_{B_R} |u-u_R|^2 dx \le \int_{B_R} |u-c|^2 dx.$$

Proof: Assume w.l.o.g. that $u_R = 0$. The statement then follows by integrating the formula

$$|u-c|^2 = |u|^2 - 2cu + c^2. \quad \square$$

2.3. Linear equations with Hölder continuous coefficients

Consider the elliptic equation

(2.20)
$$-D_\beta(a^{\alpha\beta}(x)\, D_\alpha u) = -D_\alpha f^\alpha + g.$$

Assumption (A2.1). (i) The coefficients $a^{\alpha\beta}(x)$ of the leading part are Hölder continuous functions in Ω with exponent μ, $0 < \mu < 1$, and there are numbers λ, Λ, L such that

(2.21)
$$\lambda\,|\xi|^2 \le a^{\alpha\beta}(x)\,\xi_\alpha\xi_\beta \le \Lambda\,|\xi|^2,$$

(2.22)
$$[a^{\alpha\beta}]_\mu^\Omega \le H$$

for all $x \in \Omega$, $\xi \in \mathbb{R}^n$.

(ii) The vector valued function $f = (f^1, \dots, f^n)$ belongs to the Hölder class $C^\mu(\Omega, \mathbb{R}^n)$ and $g \in L^\infty(\Omega)$.

Lemma 2.3.1. *Let* $u \in W^{1,2}(\Omega)$ *be a solution of* (2.20) *subject to the assumption* (A2.1). *Then* $u \in C^\sigma_{loc}(\Omega)$ *for all* σ, $0 < \sigma < 1$, *and for all* $x_0 \in \Omega$, *there is a radius* R_0 *and a constant* C, *which depend only on* $n, \mu, \lambda, \Lambda, H, \sigma, \mathrm{dist}(x_0, \partial\Omega)$, *such that*

(2.23)
$$\int_{B_R} |Du|^2 dx \le C R^{n-2+2\sigma}\left\{\|Du\|_{L^2(\Omega)} + [f]_\mu^\Omega + \|g\|_{L^\infty(\Omega)}\right\}^2$$

for all balls $B_R(x_0)$ *in* Ω *with* $R \le R_0$.

Proof: Let $B_{R_0} = B_{R_0}(x_0) \subset \Omega$. We freeze the coefficients $a^{\alpha\beta}(x)$ at x_0 by rewriting the equation (2.20) in the form

$$-D_\beta(a^{\alpha\beta}(x_0)\, D_\alpha u) = -D_\alpha((a^{\alpha\beta}(x_0) - a^{\alpha\beta}(x))\, D_\beta u + f^\alpha) + g$$

(2.24)
$$= -D_\alpha \tilde{f}^\alpha + g.$$

Let $v \in W^{1,2}(B_{R_0})$ be the solution of

$$-D_\beta(a^{\alpha\beta}(x_0)D_\alpha v) = 0 \quad \text{in} \quad B_{R_0}, \qquad v - u \in W_0^{1,2}(B_{R_0}). \tag{2.25}$$

Theorem 2.1.4, applied to Dv, then gives for all R, $0 < R \le R_0$,

$$\int_{B_R} |Dv|^2 dx \le C\left[\frac{R}{R_0}\right]^n \int_{B_{R_0}} |Dv|^2 dx. \tag{2.26}$$

On the other hand, from (2.24) and (2.25),

$$\int_{B_{R_0}} a^{\alpha\beta}(x_0)D_\alpha(u-v)D_\beta\eta\, dx = \int_{B_{R_0}} (\tilde{f}^\alpha - (\tilde{f}^\alpha)_{x_0,R_0})D_\alpha\eta\, dx + \int_{B_{R_0}} g\,\eta\, dx$$

for all $\eta \in W_0^{1,2}(B_{R_0})$. Let $\eta = u - v$. As in the derivation of (2.15), we employ the Poincaré inequality (1.14) to obtain

$$\int_{B_{R_0}} |D(u-v)|^2 dx \le C\left\{\int_{B_{R_0}} |\tilde{f} - (\tilde{f})_{x_0,R_0}|^2 dx + R_0^{n+2}\right\}$$

$$\le C\left\{L^2 R_0^{2\mu} \int_{B_{R_0}} |Du|^2 dx + R_0^{n+2\mu}\right\}. \tag{2.27}$$

If R_0 is sufficiently small, then this estimate can be combined with (2.26) to give

$$\int_{B_R} |Du|^2 dx \le C\left\{\left[\left[\frac{R}{R_0}\right]^n + \epsilon\right]\int_{B_{R_0}} |Du|^2 dx + R_0^{n-2+2\sigma}\right\}$$

for all σ, $0 < \sigma < 1$. Lemma 2.2.2 yields the estimate

$$\int_{B_R} |Du|^2 dx \le C\left\{\left[\frac{R}{R_0}\right]^{n-2+2\sigma} \int_{B_{R_0}} |Du|^2 dx + R^{n-2+2\sigma}\right\}$$

from which the statement follows. □

<u>Lemma 2.3.2.</u> $u \in C^{1,\sigma}_{loc}(\Omega)$ *for all* σ, $0 < \sigma < \mu$. *Furthermore there is a radius* R_0 *and a constant* C, *which depend only on* σ, $\operatorname{dist}(x_0, \partial\Omega)$ *and the data, such that for all* R, $0 < R \le R_0$,

$$\int_{B_R} |Du - (Du)_R|^2 dx \le C R^{n+2\sigma}\left\{\|Du\|_{L^2(\Omega)} + [f]_\mu^\Omega + \|g\|_{L^\infty(\Omega)}\right\}^2. \tag{2.28}$$

<u>Proof:</u> Incorporating Lemma 2.3.1, the inequality (2.27) yields

$$\int_{B_{R_0}} |D(u-v)|^2 dx \le C\left\{L^2 R_0^{2\mu}\int_{B_{R_0}} |Du|^2 dx + R_0^{n+2\mu}\right\}$$

$$\le C R_0^{n+2\sigma}$$

for all σ, $0 < \sigma < \mu$. Hence it follows that

$$\int_{B_R} |Du-(Du)_R|^2 dx \le C\left\{\left[\frac{R}{R_0}\right]^{n+2}\int_{B_{R_0}} |Du-(Du)_{R_0}|^2 dx + R_0^{n+2\sigma}\right\},$$

and the statement follows from Lemma 2.2.2. □

By the boundedness of Du, one can now prove

<u>Theorem 2.3.3.</u> *Let* $u \in W^{1,2}(\Omega)$ *be a solution of* (2.19) *subject to the assumption* (A2.1). *Then* $u \in C^{1,\mu}_{loc}(\Omega)$, *and for all* $x_0 \in \Omega$, *there is a radius* R_0 *and a constant* C *depending only on* n, μ, λ, Λ, H, $\mathrm{dist}(x_0, \partial\Omega)$, *such that*

$$\operatorname*{osc}_{B_R} Du \le C R^{\mu}\left\{\|Du\|_{L^2(\Omega)} + [f]_{\mu}^{\Omega} + \|g\|_{L^\infty(\Omega)}\right\} \tag{2.29}$$

for all balls $B_R(x_0)$ *in* Ω *with* $0 < R \le R_0$.

2.4. <u>Quasilinear elliptic systems of diagonal form</u>

Consider a quasilinear elliptic system of diagonal form with quadratic growth in the gradient of the solution mapping $u = (u^1, \ldots, u^N)$:

$$-D_\beta(a^{\alpha\beta}(x,u)\, D_\alpha u^k) = f^k(x,u,Du) \qquad (k=1,\ldots,N). \tag{2.30}$$

<u>Assumption (A2.2).</u> Suppose that $u \in W^{1,2}(\Omega, \mathbb{R}^N)$ satisfies the estimate

$$\int_\Omega |Du|^2 dx \le M. \tag{2.31}$$

<u>Assumption (A2.2').</u> Suppose further that $u \in C^0(\Omega, \mathbb{R}^N)$ such that

$$|u(x') - u(x'')| \le \omega(|x'-x''|) \qquad (x', x'' \in \Omega) \tag{2.32}$$

with a nondecreasing function $\omega: [0,\infty) \longrightarrow [0,\infty)$ which is continuous at 0 with $\omega(0) = 0$.

<u>Assumption (A2.3).</u> (i) The coefficients $a^{\alpha\beta}(x,z)$ are Hölder continuous on $\Omega \times \mathbb{R}^N$ with exponent μ, $0 < \mu < 1$, and there are positive numbers λ, Λ, H such that

(2.33)
$$\lambda |\xi|^2 \le a^{\alpha\beta}(x,z)\, \xi_\alpha \xi_\beta \le \Lambda |\xi|^2,$$

(2.34)
$$[a^{\alpha\beta}]_\mu^{\Omega\times\mathbb{R}^N} \le H$$

for all $x \in \Omega$, $z \in \mathbb{R}^N$, $\xi \in \mathbb{R}^n$.

(ii) The lower order term $f(x, u(x), Du(x))$ is a measurable $\mathbb{R}^N$–valued function on Ω satisfying, for some constants a, b,

(2.35)
$$|f(x,u(x),Du(x))| \le a\,|Du(x)|^2 + b$$

for a.e. $x \in \Omega$.

__Lemma 2.4.1.__ *Suppose that the assumptions* (A2.2, 2′, 3) *are satisfied, and let* $x_0 \in \Omega$. *Then for all* σ, $0 < \sigma < 1$, *there is a radius* R_0 *and a constant* C, *which depend only on* n, N, λ, Λ, μ, H, a, b, M, ω, *and* $\operatorname{dist}(x_0, \partial\Omega)$, *such that*

(2.36)
$$\int_{B_R(x_0)} |Du|^2\, dx \le C\, R^{n-2+2\sigma}$$

for all R, $0 < R \le R_0$.

__Proof:__ Let $R_0 < \operatorname{dist}(x_0, \partial\Omega)$. Rewrite the system (2.30) in the form

$$\begin{aligned} -D_\beta(a^{\alpha\beta}(x_0,u(x_0))\, D_\alpha u^k) &= -D_\beta((a^{\alpha\beta}(x_0,u(x_0)) - a^{\alpha\beta}(x,u(x)))\, D_\alpha u^k) + f^k \\ &= -D_\alpha \tilde{f}^{\alpha k} + f^k, \end{aligned}$$

to get the analog of (2.27):

(2.37)
$$\int_{B_{R_0}} |D(u-v)|^2\, dx \le C\Big\{\int_{B_{R_0}} |\tilde{f} - (\tilde{f})_{x_0,R_0}|^2\, dx + \int_{B_{R_0}} |f|\,|u-v|\, dx\Big\}$$

(2.38)
$$\le C\Big\{\Big[(\operatorname*{osc}_{B_{R_0}} a^{\alpha\beta}(x,u(x)))^2 + a \sup|u-v|\Big]\int_{B_{R_0}} |Du|^2\, dx + b\int_{B_{R_0}} |u-v|\, dx\Big\}.$$

Now

$$\begin{aligned} \operatorname*{osc}_{B_{R_0}} a^{\alpha\beta}(x,u(x)) &\le C\,L\,(R_0^\mu + (\operatorname*{osc}_{B_{R_0}} u)^\mu) \\ &\le C\,L\,(R_0^\mu + \omega^\mu(R_0)). \end{aligned}$$

By the maximum principle, w.l.o.g.,

$$\begin{aligned}\sup_{B_{R_0}} |u-v| &= v(x_{R_0}) - u(x_{R_0})\\ &\le \sup_{B_{R_0}} v - u(x_{R_0})\\ &\le \sup_{\partial B_{R_0}} u - u(x_{R_0})\\ &\le \operatorname{osc}_{B_{R_0}} u\\ &\le \omega(R_0).\end{aligned}$$

Hence, by employing the Poincaré inequality (1.14) to handle the latter term in (2.38), it follows that

$$\int_{B_{R_0}} |D(u-v)|^2 dx \le C\Big\{(R_0^{2\mu} + \omega^{2\mu}(R_0) + \omega(R_0)) \int_{B_{R_0}} |Du|^2 dx + R_0^{n+2}\Big\}, \tag{2.39}$$

and therefore, as in the proof of Lemma 2.3.1,

$$\int_{B_R} |Du|^2 dx \le C\Big\{\Big[\Big[\frac{R}{R_0}\Big]^n + (R_0^{2\mu} + \omega^{2\mu}(R_0) + \omega(R_0))\Big] \int_{B_{R_0}} |Du|^2 dx + R_0^{n+2}\Big\}. \tag{2.40}$$

Now let $R_1 \le \operatorname{dist}(x_0, \partial\Omega)$ be so small that

$$R_1^{2\mu} + \omega^{2\mu}(R_1) + \omega(R_1) \le \epsilon_0,$$

where $\epsilon_0 = \epsilon_0(C, n, n+2)$ is the constant from Lemma 2.2.2 and C is the one from (2.40). Then Lemma 2.2.2 yields the inequalities

$$\int_{B_R} |Du|^2 dx \le C\Big\{\Big[\frac{R}{R_0}\Big]^{n-2+2\sigma} \int_{B_{R_0}} |Du|^2 dx + R^{n-2+2\sigma}\Big\}$$

for all σ, R, R_0, $0 < \sigma < 1$, $0 < R \le R_0 \le R_1$. The statement follows. □

<u>Lemma 2.4.2.</u> *For all* σ, $0 < \sigma < \min\{\mu, 1/2\}$, *there is a radius* R_0 *and a constant* C, *which depend only on* σ, $\operatorname{dist}(x_0, \partial\Omega)$ *and the data, such that*

$$\int_{B_R} |Du - (Du)_R|^2 dx \le C R^{n+2\sigma} \tag{2.41}$$

for all R, $0 < R \le R_0$.

Proof: By (2.39), we have for all σ', $0 < \sigma' < 1$,

$$\int_{B_{R_0}} |D(u-v)|^2 dx \le C\Big\{(R_0^{2\mu} + \omega^{2\mu}(R_0) + \omega(R_0))\int_{B_{R_0}} |Du|^2 dx + R_0^{n+2}\Big\}$$

$$\le C\{(R_0^{2\mu} + R_0^{\sigma' 2\mu} + R_0^{\sigma'})R_0^{n-2+2\sigma'} + R_0^{n+2}\}.$$

Therefore

$$\int_{B_{R_0}} |D(u-v)|^2 dx \le C R_0^n (R_0^{2\sigma\mu} + R_0^{\sigma})$$

$$\le C R_0^{n+2\sigma\mu'}$$

for all σ, $0 < \sigma < 1$, $\mu' = \min\{\mu, 1/2\}$, from which, as in the proof of Lemma 2.3.2,

$$\int_{B_R} |Du-(Du)_R|^2 dx \le C\Big\{\Big[\frac{R}{R_0}\Big]^{n+2}\int_{B_{R_0}} |Du-(Du)_{R_0}|^2 dx + R_0^{n+2\sigma\mu'}\Big\}.$$

This implies the stated inequality (2.41). □

Theorem 2.4.3. *Let* $u \in W^{1,2}(\Omega)$ *be a solution to* (2.30) *with the assumptions* (A2.2, 2′, 3) *satisfied. Then* $u \in C^{1,\mu}_{loc}(\Omega)$, *and for each* $x_0 \in \Omega$, *there is a radius* R_0 *and a constant* C, *which depend only on* n, N, λ, Λ, μ, H, a, b, M, ω *and* $\operatorname{dist}(x_0, \partial\Omega)$, *such that*

$$\operatorname*{osc}_{B_R(x_0)} Du \le C R^{\mu} \tag{2.42}$$

for all R, $0 < R \le R_0$.

Proof: We go back to the inequality (2.37) and use the boundedness of Du to estimate

$$\int_{B_{R_0}} |D(u-v)|^2 dx \le C\Big\{\int_{B_{R_0}} |\tilde{f}-(\tilde{f})_{x_0,R_0}|^2 dx + \int_{B_{R_0}} |f|\,|u-v|\,dx\Big\}$$

$$\le C\Big\{(\operatorname*{osc}_{B_{R_0}} a^{\alpha\beta})^2\int_{B_{R_0}} |Du|^2 dx + \int_{B_{R_0}} |u-v|\,dx\Big\}$$

$$\le C\Big\{R_0^{n+2\mu} + \frac{\epsilon}{R_0^2}\int_{B_{R_0}} |u-v|^2 dx + \frac{R_0^{n+2}}{\epsilon}\Big\},$$

and hence

$$\int_{B_{R_0}} |D(u-v)|^2 dx \le C R_0^{n+2\mu}.$$

The statement follows. □

Theorem 2.4.3 actually follows directly from the linear theory, Theorem 2.3.3, via Lemma 2.4.2, because once the boundedness of Du is known, then the quasilinear system (2.30) splits into two linear equations (from the estimates point of view).

<u>Theorem 2.4.4.</u> *Suppose that* $\Omega \subset \mathbb{R}^2$ *and that* $u : \Omega \longrightarrow u(\Omega)$ *is a homeomorphic solution to* (2.30) *of class* $W^{1,2}(\Omega, \mathbb{R}^2)$ *subject to the assumptions* (A2.2, 3).
Then $u \in C^{1,\mu}_{loc}(\Omega)$, *and for each* $x_0 \in \Omega$, *there is a radius* R_0 *and a constant* C, *which depend only on* λ, Λ, μ, H, a, b, M *and* $\operatorname{dist}(x_0, \partial\Omega)$, *such that*

(2.43)
$$\operatorname*{osc}_{B_R(x_0)} Du \le C R^{\mu}$$

for all R, $0 < R \le R_0$.

<u>Proof:</u> We only have to verify the assumption (A2.2′), which is true because of the Courant–Lebesgue lemma, Lemma 1.6.1:

$$\operatorname*{osc}_{B_R(x_0)} u \le 2 \sqrt{\frac{\pi M}{\log 1/R}} = \omega(R)$$

if $B_{\sqrt{R}} \subset \Omega$, $0 < R < 1$, as required. □

Chapter 3. REGULARITY FOR MONGE–AMPÈRE EQUATIONS

3.1. Ellipticity and convexity

Let Ω be an open set in the x, y–plane.

Definition 3.1.1. A partial differential equation of the form

$$F(x, y, z, p, q, r, s, t) = 0 \tag{3.1}$$

($p = z_x$, $q = z_y$, $r = z_{xx}$, $s = z_{xy}$, $t = z_{yy}$) is *elliptic* for a given solution $z \in C^{1,1}(\Omega)$, if the characteristic form is positive definite, i.e.,

$$F_r \xi^2 + F_s \xi\eta + F_t \eta^2 > 0 \tag{3.2}$$

for all $(\xi, \eta) \in \mathbb{R}^2$, $\xi^2 + \eta^2 > 0$, and $(x, y) \in \Omega$. Here $F_r = F_r(x, y, z(x, y), \dots, t(x, y)), \dots$, $F_t = F_t(\dots)$. The function F is assumed to be differentiable with respect to the variables r, s, t and continuous with respect to x, y, z, p, q.

The equation (3.1) is *uniformly elliptic* if there are positive constants λ, Λ such that

$$\lambda(\xi^2 + \eta^2) \le F_r \xi^2 + F_s \xi\eta + F_t \eta^2 \le \Lambda(\xi^2 + \eta^2) \tag{3.3}$$

for $(\xi, \eta) \in \mathbb{R}^2$, $(x, y) \in \Omega$.

Remark 3.1.2. F is elliptic iff the discriminant is positive,

$$\Delta(x, y, z(x,y), \dots, t(x,y)) = F_r F_t - \tfrac{1}{4} F_s^2 > 0, \tag{3.4}$$

and if $F_r > 0$.

Proof: Suppose that (3.4) holds. Then $F_r > 0$ and $F_t > 0$, and

$$F_r \xi^2 + F_s \xi\eta + F_t \eta^2 = F_r \left[\left[\xi + \frac{1}{2} \frac{F_s}{F_r} \eta \right]^2 + \frac{1}{F_r^2} (F_r F_t - \tfrac{1}{4} F_s^2) \eta^2 \right]$$
$$\ge \frac{\Delta}{F_r} \eta^2 .$$

It follows that

$$F_r \xi^2 + F_s \xi\eta + F_t \eta^2 \ge \frac{\Delta}{2} \left[\frac{1}{F_t} \xi^2 + \frac{1}{F_r} \eta^2 \right] > 0 \tag{3.5}$$

for $(\xi, \eta) \in \mathbb{R}^2$, $\xi^2 + \eta^2 > 0$, and $(x, y) \in \Omega$. □

Consider a solution $z \in C^{1,1}(\Omega)$ of the Monge–Ampère equation

$$rt - s^2 = f(x,y,z,p,q). \tag{3.6}$$

Lemma 3.1.3. *The equation* (3.6) *is elliptic for* z *iff* $t > 0$ *and*

$$f(x,y,z(x,y),\dots,q(x,y)) > 0 \tag{3.7}$$

for $(x,y) \in \Omega$. *If* $z \in C^2(\Omega)$, *then ellipticity is true iff* z *is strictly convex.*

Proof: This is obvious by

$$\Delta = F_r F_t - \tfrac{1}{4} F_s^2 = rt - s^2 = f. \ \square$$

Lemma 3.1.4. *Let the solution* $z \in C^{1,1}(\Omega)$ *of* (3.6) *satisfy the estimates*

$$|D^2 z| \le K, \tag{3.8}$$

$$f(x,y,z(x,y),\dots,q(x,y)) \ge \frac{1}{c} \tag{3.9}$$

for $(x,y) \in \Omega$, *and suppose that* $t > 0$. *The equation* (3.6) *is then uniformly elliptic in* Ω *with*

$$\begin{aligned} \frac{1}{2cK}(\xi^2 + \eta^2) &\le F_r \xi^2 + F_s \xi\eta + F_t \eta^2 \\ &= t\xi^2 - 2s\xi\eta + r\eta^2 \\ &\le 2K(\xi^2 + \eta^2). \end{aligned} \tag{3.10}$$

Furthermore

$$r, t \ge \frac{1}{cK}. \tag{3.11}$$

Proof: The inequality (3.5) yields

$$\begin{aligned} F_r \xi^2 + F_s \xi\eta + F_t \eta^2 &\ge \frac{\Delta}{2}\left[\frac{1}{F_t}\xi^2 + \frac{1}{F_r}\eta^2\right] \\ &= \frac{f}{2}\left[\frac{1}{r}\xi^2 + \frac{1}{t}\eta^2\right] \\ &\ge \frac{1}{2cK}(\xi^2 + \eta^2), \end{aligned}$$

the only not so obvious inequality. $\square$

3.2. A Legendre like transformation

Let $z \in C^{1,1}$ be a solution of the elliptic Monge–Ampère equation

$$rt - s^2 = f(x,y) > 0 \tag{3.12}$$

in the disc $B = B_{R_0}(x_0, y_0)$. I.e., assume w.l.o.g. that $t > 0$. Consider the transformation T:

$$\begin{aligned} u &= x, \\ v &= q = z_y(x,y) \end{aligned} \tag{3.13}$$

for $(x,y) \in B$, when the equation (3.12) is uniformly elliptic, i.e. require

Assumption (A3.1). For $(x,y) \in B$ let

$$|D^2 z| \le K, \tag{3.14}$$

$$f(x,y) \ge \frac{1}{c}. \tag{3.15}$$

Lemma 3.2.1. *For* $(x', y'), (x'', y'') \in B$ *the following dilation estimates hold*:

$$(u'-u'')^2 + (v'-v'')^2 \le \gamma_1^2((x'-x'')^2 + (y'-y'')^2), \tag{3.16}$$

$$(x'-x'')^2 + (y'-y'')^2 \le \gamma_2^2((u'-u'')^2 + (v'-v'')^2). \tag{3.17}$$

Here $u' = u(x', y'), \ldots, v'' = v(x'', y'')$, *and* $\gamma_1 = \gamma_1(K)$, $\gamma_2 = \gamma_2(c, K)$ *are constants* ≥ 1.

Corollary 3.2.2. T *is a bi-Lipschitz map from* B *onto its image* $T(B)$ *with Jacobian*

$$J_T(x,y) = u_x v_y - u_y v_x = t \ge \frac{1}{cK}. \tag{3.18}$$

The inclusions

$$T(B_{R/\gamma_1}(x_0, y_0)) \subset B_R(u_0, v_0), \tag{3.19}$$

$$B_{R/\gamma_2}(u_0, v_0) \subset T(B_R(x_0, y_0)) \tag{3.20}$$

hold for all R, $0 < R \le R_0$.

Proof of Lemma 3.2.1: The Lipschitz continuity of q implies

$$v' - v'' = q(x', y') - q(x'', y'')$$

$$= \int_0^1 \frac{d}{d\tau} q((x'', y'') + \tau(x' - x'', y' - y''))\, d\tau$$

$$= \int_0^1 \{s(\dots)(x' - x'') + t(\dots)(y' - y'')\}\, d\tau,$$

from which the inequality (3.16) follows. (3.17) is shown by estimating

$$|v' - v''| \geq -K\,|x' - x''| + \frac{1}{cK}\,|y' - y''|,$$

and therefore

$$\frac{1}{cK}\,|y' - y''| \leq K\,|u' - u''| + |v' - v''|$$

as required. □

<u>Lemma 3.2.3.</u> *The function* $y(u,v) \in C^{0,1}(T(B))$ *is a weak solution of the equation*

(3.21)
$$y_{uu} + (f y_v)_v = 0,$$

which is trivially satisfied by $x(u,v)$, *and*

(3.22)
$$\begin{bmatrix} x_u & x_v \\ y_u & y_v \end{bmatrix} = \begin{bmatrix} 1 & 0 \\ -s/t & 1/t \end{bmatrix}.$$

<u>Proof:</u> Compute

$$\begin{bmatrix} u_x & u_y \\ v_x & v_y \end{bmatrix} = \begin{bmatrix} 1 & 0 \\ s & t \end{bmatrix},$$

from which (3.22) follows. Let $\eta \in C_0^\infty(T(B))$, then

$$\int_{T(B)} (y_u \eta_u + f y_v \eta_v)\, du\, dv = \int_B \{(-s/t)(\eta_x + (-s/t)\eta_y) + f(1/t)^2 \eta_y\}\, t\, dx\, dy$$

$$= \int_B (-s\,\eta_x + r\,\eta_y)\, dx\, dy$$

$$= 0$$

by approximation. □

<u>Lemma 3.2.4.</u> *Assume only that*

(3.23)
$$|Dz| \leq K_1, \qquad f(x,y) \geq \frac{1}{c}.$$

Then

(3.24)
$$\int_{T(B)} |Dy|^2\, du\, dv \leq C(c, K_1)\, R_0.$$

Proof: By (3.22),

$$\begin{aligned}\int_{T(B)} y_u^2\,du\,dv &= \int_B \frac{s^2}{t}\,dx\,dy \\ &= \int_B \left(r - \frac{f}{t}\right) dx\,dy \\ &\le \int_B r\,dx\,dy - \frac{1}{c}\int_B \frac{1}{t}\,dx\,dy \\ &= \int_B r\,dx\,dy - \frac{1}{c}\int_{T(B)} y_v^2\,du\,dv .\end{aligned}$$

A general divergence theorem therefore gives

$$\begin{aligned}\int_{T(B)} |Dy|^2\,du\,dv &\le C\int_{\partial B} |p|\,ds \\ &\le 2\pi C K_1 R_0\end{aligned}$$

as required. □

Lemma 3.2.5. *If*

(3.25) $$q(x,y) \ge 1,$$

then the function $z(u,v) \in C^{0,1}(T(B))$ *is a weak solution of*

(3.26) $$z_{uu} + (f z_v)_v - \frac{2f}{v} z_v = 0,$$

and

(3.27) $$\begin{bmatrix} z_u \\ z_v \end{bmatrix} = \begin{bmatrix} 1 & -s/t \\ 0 & 1/t \end{bmatrix} \begin{bmatrix} p \\ q \end{bmatrix}.$$

Proof:

$$\begin{bmatrix} p \\ q \end{bmatrix} = \begin{bmatrix} 1 & s \\ 0 & t \end{bmatrix} \begin{bmatrix} z_u \\ z_v \end{bmatrix},$$

which implies (3.27). For any $\eta \in C_0^\infty(T(B))$:

$$\begin{aligned}\int_{T(B)} (z_u \eta_u + f z_v \eta_v)\,du\,dv &= \int_B \{(p + (-s/t)q)(\eta_x + (-s/t)\eta_y) + f(1/t)^2 q\,\eta_y\}\,t\,dx\,dy \\ &= \int_B \{(pt - sq)\,\eta_x + (qr - ps)\,\eta_y\}\,dx\,dy\end{aligned}$$

(3.28) $$= -2\int_B f\,\eta\,dx\,dy,$$

again by approximation. □

Let us remark that the relation (3.28) means that the Monge–Ampère equation (3.12) is of divergence structure. It is this observation that leads to the current approach.

3.3. The equation $rt - s^2 = f$

Assumption (A3.2). Let $f = f(x, y, z, p, q)$ be a positive function of class $C^{\mu}(\Omega \times \mathbb{R}^3)$ for some μ, $0 < \mu < 1$, and suppose that there are positive constants a, b, c such that

(3.29) $$\frac{1}{c} \leq f(x, y, z, p, q) \leq a$$

(3.30) $$[f]_{\mu}^{\Omega \times \mathbb{R}^3} \leq b.$$

Theorem 3.3.1. *Let* $z \in C^{1,1}(\Omega)$ *be a solution of the Monge–Ampère equation*

(3.31) $$rt - s^2 = f(x, y, z, p, q)$$

with

(3.32) $$||z||_{C^{1,1}(\Omega)} \leq K.$$

Suppose that the assumption (A3.2) *is satisfied.*

Then $z \in C^{2,\mu}_{loc}(\Omega)$, *and for* $\Omega' \subset\subset \Omega$, *there is a constant* C *which depends only on* μ, a, b, c, K, $\mathrm{dist}(\Omega', \partial\Omega)$ *such that*

(3.33) $$[D^2 z]_{\mu}^{\Omega'} \leq C.$$

Proof: Assume w.l.o.g. that $t > 0$. In view of the bound (3.32), we can also w.l.o.g. assume that f only depends on (x, y). Let $B = B_{R_0}(x_0, y_0) \subset \Omega$, and consider the Legendre–like transformation T (3.13),

$$u = x,$$

$$v = q(x, y)$$

in B. The function $y(u, v)$ satisfies the equation (3.21),

$$y_{uu} + (f y_v)_v = 0$$

in T(B), hence in $B_{R_0/\gamma_2}(u_0, v_0)$ by the inclusion (3.20). Also

$$|Dy| \leq C(c, K)$$

because of (3.22) and (3.11).

We can therefore apply the linear theory, Theorem 2.3.3, to obtain $y \in C^{1,\mu}_{loc}(B_{R_0/\gamma_2}(u_0,v_0))$ together with the estimate

$$\operatorname*{osc}_{B_R(u_0,v_0)} Du \le C R^\mu$$

for each R with $R \le R_1 = \min\{R_0/\gamma_2, R_2\}$ where $R_2 = R_2(\mu, a, b, c, K)$. The representations (3.22) yield the regularity $t, s \in C^\mu_{loc}(B_{R_0/\gamma_2}(u_0,v_0))$ and

$$\operatorname*{osc}_{B_R(u_0,v_0)} \{t,s\} \le C R^\mu$$

for $R \le R_1$. Hence $t, s \in C^\mu_{loc}(B_{R_0/\gamma_1\gamma_2}(x_0,y_0))$ and

$$\operatorname*{osc}_{B_R(x_0,y_0)} \{t,s\} \le C R^\mu$$

for $R \le R_1/\gamma_1$ by the inclusion (3.19). The differential equation (3.31) gives the regularity $D^2z \in C^\mu_{loc}(B_{R_0/\gamma_1\gamma_2}(x_0,y_0))$ and the estimate

$$\operatorname*{osc}_{B_R(x_0,y_0)} D^2z \le C R^\mu$$

for $R \le R_1/\gamma_1$ as required. □

3.4. The general Monge–Ampère equation

Consider a solution $z \in C^{1,1}(\Omega)$ of an elliptic Monge–Ampère equation of the general form

$$Ar + 2Bs + Ct + (rt - s^2) = E \tag{3.34}$$

or equivalently,

$$(r+C)(t+A) - (s-B)^2 = \Delta, \tag{3.35}$$

where

$$\Delta = AC - B^2 + E \tag{3.36}$$

is the discriminant. In view of (3.4), ellipticity of (3.34) means that $\Delta > 0$ in Ω and $t+A > 0$.

<u>Assumption (A3.3).</u> Suppose that $z \in C^{1,1}(\Omega)$ satisfies the estimate

$$\|z\|_{C^{1,1}(\Omega)} \le K. \tag{3.37}$$

<u>Assumption (A3.4).</u> The coefficients A, B, C and the nonhomogeneity E are μ–Hölder continuous functions with respect to the five variables x, y, z, p, q such that

$$\sup_{\Omega\times\mathbb{R}^3} \{ |A|, |B|, |C|, |E| \} \le a, \tag{3.38}$$

$$[A, B, C, E]_{\mu}^{\Omega\times\mathbb{R}^3} \le b, \tag{3.39}$$

$$\Delta = AC - B^2 + E \ge \tfrac{1}{c}. \tag{3.40}$$

These requirements are needed only on the solution surface $\{(x,y,z(x,y),p(x,y),q(x,y)) \mid (x,y) \in \Omega\}$. We can w.l.o.g. assume that the coefficients A, B, C and E depend only on x, y and freeze A, B, C. That is, for $(x_0, y_0) \in \Omega$ we set

$$A_0 = A(x_0, y_0), \ldots, E_0 = E(x_0, y_0), \quad \Delta_0 = \Delta(x_0, y_0). \tag{3.41}$$

<u>Lemma 3.4.1.</u> *Suppose that the assumptions* (A3.3,4) *are satisfied. The function*

$$\tilde{z}(x,y) = z(x,y) + \tfrac{1}{2}(C_0 (x-x_0)^2 - 2 B_0 (x-x_0)(y-y_0) + A_0 (y-y_0)^2), \tag{3.42}$$

$(x,y) \in \Omega$, *solves the Monge–Ampère equation*

$$\tilde{r}\,\tilde{t} - \tilde{s}^2 = \tilde{f}(x,y), \tag{3.43}$$

where

$$\tilde{f}(x,y) = \Delta_0 + (A_0 - A)\,r + 2(B_0 - B)\,s + (C_0 - C)\,t + (E - E_0). \tag{3.44}$$

Furthermore there exists a radius R_0, *which depends only on* μ, b, c, K *and* $\mathrm{dist}((x_0,y_0), \partial\Omega)$, *such that the equation* (3.43) *is uniformly elliptic in* $B_{R_0}(x_0, y_0)$, *i.e., the inequality*

$$\tilde{f}(x,y) \ge \frac{1}{2c} = \frac{1}{\tilde{c}} \tag{3.45}$$

holds for $(x,y) \in B_{R_0}$. *Naturally, of course,*

$$\|D^2 \tilde{z}\|_{C^{1,1}(B_{R_0})} \le \tilde{K} = \tilde{K}(a, K), \tag{3.46}$$

$$\tilde{r}, \tilde{t} \ge \frac{1}{\tilde{c}\tilde{K}}. \tag{3.47}$$

<u>Proof:</u> The equation (3.35) can be expressed in the form

$$(r+C_0)(t+A_0)-(s-B_0)^2=\tilde{f}(x,y),$$

which is equivalent to (3.43). The ellipticity follows by estimating

$$\begin{aligned}\tilde{f}(x,y) &\geq \frac{1}{c}-5bK\{(x-x_0)^2+(y-y_0)^2\}^{\mu/2}\\ &\geq \frac{1}{2c}\end{aligned}$$

if $(x,y)\in B_{R_0}(x_0,y_0)$,

$$R_0=\min\left\{1/\sqrt[\mu]{10bcK},\ \operatorname{dist}((x_0,y_0),\partial\Omega)\right\},$$

as required. □

<u>Lemma 3.4.2.</u> *Consider the variable transformation* $\tilde{T}$:

(3.48)
$$\begin{aligned}u&=x,\\ v&=\tilde{q}(x,y)\end{aligned}$$

for $(x,y)\in B_{R_0}(x_0,y_0)$. *Then* $y(u,v)\in C^{0,1}(B_{R_0/\gamma_2}(u_0,v_0))$ *satisfies the estimate*

(3.49)
$$\int_{B_R(u_0,v_0)}|Dy-(Dy)_R|^2\,du\,dv\leq\frac{C}{R_0^{2\mu}}R^{2+2\mu}$$

for all R, $0<R<R_0/\gamma_2$. *Here* $\gamma_2=\gamma_2(a,c,K)$ *is the dilation constant from Lemma* 3.2.1, *and the constant* C *depends only on the data.*

<u>Proof:</u> By Lemma 3.2.3, $y(u,v)\in C^{0,1}(B_{R_0/\gamma_2}(u_0,v_0))$ solves the equation

$$y_{uu}+(\tilde{f}\,y_v)_v=0,$$

which can be rewritten in the form

(3.50)
$$\begin{aligned}y_{uu}+(\Delta_0 y_v)_v&=\tilde{g}_v\\ &=\left[\{(A-A_0)r+2(B-B_0)s+(C-C_0)t+(E_0-E)\}\frac{1}{\tilde{t}}\right]_v.\end{aligned}$$

The dilation constants γ_1, γ_2 from Lemma 3.2.1 depend only on a, c, K through $\tilde{c}=2c$, $\tilde{K}(a,K)$. Note also that

(3.51)
$$\begin{bmatrix}x_u & x_v\\ y_u & y_v\end{bmatrix}=\begin{bmatrix}1 & 0\\ -\tilde{s}/\tilde{t} & 1/\tilde{t}\end{bmatrix}.$$

The Campanato technique as described in Section 2.2 can now be applied to (3.50): As in the proof of Theorem 2.2.1, or, more precisely, by the error estimate (2.27) for equation (2.24), we obtain the inequalities

$$\int_{B_\rho} |Dy-(Dy)_\rho|^2 du\,dv \le C\Big\{\Big[\frac{\rho}{R}\Big]^4 \int_{B_R} |Dy-(Dy)_R|^2 du\,dv + \int_{B_R} |\tilde{g}-(\tilde{g})_R|^2 du\,dv\Big\}$$

for all $\rho \le R \le R_0/\gamma_2$. Here all discs are centered at (u_0, v_0). The dilation inequalities (3.16), (3.17) are used to estimate

$$\int_{B_\rho} |Dy-(Dy)_\rho|^2 du\,dv \le C\Big\{\Big[\frac{\rho}{R}\Big]^4 \int_{B_R} |Dy-(Dy)_R|^2 du\,dv + R^{2+2\mu}\Big\}.$$

The iteration lemma, Lemma 2.2.2, yields for all $\rho < R \le R_0/\gamma_2$,

$$\begin{aligned}\int_{B_\rho} |Dy-(Dy)_\rho|^2 du\,dv &\le C\Big\{\Big[\frac{\rho}{R}\Big]^{2+2\mu} \int_{B_R} |Dy-(Dy)_R|^2 du\,dv + \rho^{2+2\mu}\Big\}\\ &\le \frac{C}{R^{2\mu}}\rho^{2+2\mu},\end{aligned}$$

from which the statement follows. □

Lemma 3.4.2. does not immediately imply Hölder continuity, because the center (u_0, v_0) is fixed. The variables u, v, and therefore the function y(u, v), depend on the point (u_0, v_0).

<u>**Theorem 3.4.3.**</u> *Let* $z \in C^{1,1}(\Omega)$ *be a solution of* (3.34) *subject to the assumptions* (A3.3, 4). *Then* $z \in C^{2,\mu}_{loc}(\Omega)$, *and for each* Ω', $\Omega' \subset\subset \Omega$, *there is a constant* C *which depends only on* μ, a, b, c, K, $\mathrm{dist}(\Omega', \partial\Omega)$ *such that*

(3.52) $$[D^2 z]_\mu^{\Omega'} \le C.$$

<u>**Proof:**</u> Re-introducing the variables x, y yields, by (3.51),

$$\begin{aligned} y_v-(y_v)_R &= \frac{1}{\tilde{t}} - \frac{1}{|B_R|}\int_{B_R} \frac{1}{\tilde{t}} du\,dv\\ &= \frac{1}{\tilde{t}}\frac{|T^{-1}(B_R)|}{|B_R|}\Big[\frac{|B_R|}{|T^{-1}(B_R)|} - \tilde{t}\Big]\\ &= \frac{1}{\tilde{t}}\frac{|T^{-1}(B_R)|}{|B_R|}(t_{;R} - t)\end{aligned}$$

a. e., where

$$g_{;R} = g_{(x_0,y_0);R} = \frac{1}{|T^{-1}(B_R)|}\int_{T^{-1}(B_R)} g(x,y)\,dx\,dy.$$

Inserting this into (3.49) gives the estimate

$$\int_{B_{R/\gamma_1}(x_0,y_0)} |t - t_{;R}|^2 \, dx\, dy \le \frac{C}{R_0^{2\mu}} R^{2+2\mu} \tag{3.53}$$

for all R, $0 < R \le R_0/\gamma_2$.

Similarly,

$$y_u - (y_u)_R = \frac{1}{|B_R|}\int_{B_R} \frac{\tilde{s}}{\tilde{t}} \, du\, dv - \frac{\tilde{s}}{\tilde{t}}$$

$$= \frac{1}{\tilde{t}} \frac{|T^{-1}(B_R)|}{|B_R|} (\tilde{t}\,\tilde{s}_{;R} - \tilde{s}\,\tilde{t}_{;R}).$$

On estimating

$$|s - s_{;R}| \le \left|\tilde{s} - \frac{\tilde{s}_{;R}}{\tilde{t}_{;R}}\tilde{t}\right| + \left|\frac{\tilde{s}_{;R}}{\tilde{t}_{;R}}\tilde{t} - \tilde{s}_{;R}\right|$$

$$\le C\{\,|\tilde{s}\,\tilde{t}_{;R} - \tilde{s}_{;R}\,\tilde{t}| + |t - t_{;R}|\,\},$$

we therefore conclude, again using (3.49), that

$$\int_{B_{R/\gamma_1}} |s - s_{;R}|^2 \, dx\, dy \le \frac{C}{R_0^{2\mu}} R^{2+2\mu} \tag{3.54}$$

for all R, $0 < R \le R_0/\gamma_2$.

The growth estimates (3.53) and (3.54) yield the Hölder continuity of s and t by virtue of the Campanato criterion, Theorem 1.4.2. The derivative r is Hölder continuous by the differential equation (3.34), as stated. □

Chapter 4. FUNCTION THEORY OF ELLIPTIC EQUATIONS

In this chapter, which will be continued later, we prepare for the investigation of homeomorphic solutions of certain quasilinear elliptic systems.

It is convenient to introduce complex notation $z = x + iy = (x, y)$ and employ the Wirtinger operators

$$\frac{\partial}{\partial z} = \frac{1}{2}\left[\frac{\partial}{\partial x} - i\frac{\partial}{\partial y}\right], \tag{4.1}$$

$$\frac{\partial}{\partial \bar{z}} = \frac{1}{2}\left[\frac{\partial}{\partial x} + i\frac{\partial}{\partial y}\right]. \tag{4.2}$$

The Laplace operator can then be rewritten in the form

$$\Delta = 4\frac{\partial^2}{\partial \bar{z}\,\partial z}. \tag{4.3}$$

Furthermore, if $w = u + iv = (u, v)$ is a C^1-function in the x, y-plane,

$$|Dw|^2 = u_x^2 + u_y^2 + v_x^2 + v_y^2$$

$$= 2(|w_z|^2 + |w_{\bar{z}}|^2), \tag{4.4}$$

$$Jw = u_x v_y - u_y v_x$$

$$= |w_z|^2 - |w_{\bar{z}}|^2, \tag{4.5}$$

and the complex form of the divergence theorem reads as

$$\int_{\partial\Omega} w\,dz = 2i\int_{\Omega} w_{\bar{z}}\,dx\,dy, \tag{4.6}$$

$$\int_{\partial\Omega} w\,d\bar{z} = -2i\int_{\Omega} w_z\,dx\,dy. \tag{4.7}$$

4.1. Hadamard's integral estimates

The following estimates were first derived by Hadamard (compare Goursat [GO], pp. 360–364):

Lemma 4.1.1. *Let* $z_1, z_2 \in B_R = B_R(z_0)$, $|z_1 - z_2| \le 1$, *and let* $\alpha, \beta < 2$. *Then*

$$\int_{B_R} \frac{dx\,dy}{|z-z_1|^{\alpha}|z-z_2|^{\beta}} \le \begin{cases} C(\alpha,\beta,R) & (\alpha+\beta<2) \\ C(\alpha,\beta,R) + 8\pi \log 1/|z_1-z_2| & (\alpha+\beta=2). \\ C(\alpha,\beta)\,|z_1-z_2|^{2-\alpha-\beta} & (\alpha+\beta>2) \end{cases} \tag{4.8}$$

Proof: If $\rho = |z_1 - z_2|$, then

$$\int_{B_R} \frac{dx\,dy}{|z-z_1|^\alpha |z-z_2|^\beta} \le \int_{B_{2\rho}(z_1)} \frac{dx\,dy}{|z-z_1|^\alpha |z-z_2|^\beta} + \int_{B_{4R}(z_1)\setminus B_{2\rho}(z_1)} \frac{dx\,dy}{|z-z_1|^\alpha |z-z_2|^\beta}$$
$$= I_1 + I_2 .$$

But $|z-z_1| \ge 2\rho$ implies

$$\begin{aligned} |z-z_2| &\ge |z-z_1| - |z_1-z_2| \\ &= |z-z_1| - \rho \\ &\ge \tfrac{1}{2}|z-z_1| , \end{aligned}$$

and therefore

$$\begin{aligned} I_2 &\le 4 \int_{B_{4R}(z_1)\setminus B_{2\rho}(z_1)} \frac{dx\,dy}{|z-z_1|^{\alpha+\beta}} \\ &= 8\pi \int_{2\rho}^{4R} r^{1-\alpha-\beta}\,dr \end{aligned}$$

$$\text{(4.9)} \qquad \le \begin{cases} \dfrac{8\pi}{2-\alpha-\beta} (4R)^{2-\alpha-\beta} & (\alpha+\beta < 2) \\ 8\pi \log \dfrac{2R}{|z_1-z_2|} & (\alpha+\beta = 2) . \\ \dfrac{8\pi}{\alpha+\beta-2} (2|z_1-z_2|)^{2-\alpha-\beta} & (\alpha+\beta > 2) \end{cases}$$

Furthermore, by substituting $z = z_1 + 2\rho\zeta$,

$$\begin{aligned} I_1 &= \int_{B_1(0)} \frac{(2\rho)^2\,d\xi\,d\eta}{(2\rho)^\alpha |\zeta|^\alpha\, |2\rho\zeta + z_1 - z_2|^\beta} \\ &= (2\rho)^{2-\alpha-\beta} \int_{B_1(0)} \frac{d\xi\,d\eta}{|\zeta|^\alpha |\zeta + e^{i\vartheta}|^\beta} \\ &= C(\alpha,\beta)\, |z_1-z_2|^{2-\alpha-\beta} , \end{aligned}$$

if $z_1 \ne z_2$. The stated estimates follow. □

4.2. The nonhomogeneous Cauchy–Riemann system

Lemma 4.2.1. *Suppose that* Ω *is a smooth domain. Let* $w \in C^1_{loc}(\Omega) \cap C^0(\bar\Omega)$ *be a solution of the nonhomogeneous Cauchy–Riemann system*

$$\text{(4.10)} \qquad w_{\bar z} = f \in C^0(\bar\Omega) .$$

Then for $z \in \Omega$ *the following representation holds:*

$$w(z) = \frac{1}{2\pi i}\int_{\partial\Omega} \frac{w(\zeta)}{\zeta - z} d\zeta - \frac{1}{\pi}\int_{\Omega} \frac{f(\zeta)}{\zeta - z} d\xi\, d\eta. \tag{4.11}$$

Proof: Let $\Omega_\epsilon = \Omega \setminus B_\epsilon(z)$. By multiplying $w_{\bar\zeta} = f$ with

$$h(\zeta) = \frac{1}{\zeta - z},$$

one obtains

$$\begin{aligned}\int_{\Omega_\epsilon} f h\, d\xi\, d\eta &= \int_{\Omega_\epsilon} (w h)_{\bar\zeta}\, d\xi\, d\eta \\ &= \frac{1}{2i}\int_{\partial\Omega_\epsilon} w h\, d\zeta \\ &= \frac{1}{2i}\int_{\partial\Omega} w h\, d\zeta - \frac{w(\tilde z)}{2i}\int_{\partial B_\epsilon} \frac{d\zeta}{\zeta - z}\end{aligned}$$

for some $\tilde z \in \partial B_\epsilon$, by the Gauß–Green theorem (4.6). Since

$$\int_{\partial B_\epsilon} \frac{d\zeta}{\zeta - z} = 2\pi i$$

$(\zeta = z + \epsilon e^{i\vartheta})$, it follows that

$$\int_{\Omega} \frac{f(\zeta)}{\zeta - z} d\xi\, d\eta = \frac{1}{2i}\int_{\partial\Omega} \frac{w(\zeta)}{\zeta - z} d\zeta - \pi w(\tilde z)$$

as required. □

Lemma 4.2.2. *Suppose that* $\Omega \subset B_R$, $B_R = B_R(0)$, *and let* $f \in L^\infty(\Omega)$ *be such that*

$$\sup_{\Omega} |f| \le a. \tag{4.12}$$

Then the integral

$$g(z) = Tf(z) = T_\Omega f(z) = -\frac{1}{\pi}\int_{\Omega} \frac{f(\zeta)}{\zeta - z} d\xi\, d\eta \tag{4.13}$$

satisfies for all μ, $0 < \mu < 1$, *the estimates*

$$\sup_{B_R} |g| \le 4aR, \tag{4.14}$$

$$[g]_\mu^{B_R} \le C(\mu, a, R). \tag{4.15}$$

<u>**Proof:**</u> For $z_1, z_2 \in B_R$, $z_1 \neq z_2$, let $\rho = |z_1 - z_2|$. Then

$$|g(z_1) - g(z_2)| \leq \frac{a}{\pi}\int_{B_R} \left|\frac{1}{\zeta - z_1} - \frac{1}{\zeta - z_2}\right| d\xi\, d\eta$$

$$\leq \frac{a}{\pi}\int_{B_{2\rho}(z_1)} \frac{d\xi\, d\eta}{|\zeta - z_1|} + \frac{a}{\pi}\int_{B_{2\rho}(z_1)} \frac{d\xi\, d\eta}{|\zeta - z_2|}$$

$$+ \frac{a}{\pi}\int_{B_R \cap \{|\zeta - z_1| \geq 2\rho\}} \left|\frac{1}{\zeta - z_1} - \frac{1}{\zeta - z_2}\right| d\xi\, d\eta$$

$$\leq \frac{a}{\pi}\int_{B_{2\rho}(z_1)} \frac{d\xi\, d\eta}{|\zeta - z_1|} + \frac{a}{\pi}\int_{B_{3\rho}(z_2)} \frac{d\xi\, d\eta}{|\zeta - z_2|}$$

$$+ \frac{a}{\pi}|z_1 - z_2| \int_{2\rho \leq |\zeta - z_1| \leq 4R} \frac{d\xi\, d\eta}{|\zeta - z_1|\,|\zeta - z_2|}.$$

The statement follows from Lemma 4.1.1, or, more precisely, from (4.9), because $(\mu-1)\log\rho \leq \rho^{\mu-1}$ $(0 < \rho \leq 1)$ for all μ, $0 < \mu < 1$. □

<u>Proposition 4.2.3.</u> *Let* $f \in L^\infty(\Omega)$. *Then* $g = Tf$ *satisfies the equation* (4.10) *in the weak sense, i.e., for all* $\eta \in C_0^1(\Omega)$,

$$\int_\Omega g\, \eta_{\bar{z}}\, dx\, dy = -\int_\Omega f\, \eta\, dx\, dy. \tag{4.16}$$

<u>**Proof:**</u> By Lemma 4.2.1,

$$\eta(z) = -\frac{1}{\pi}\int_\Omega \frac{\eta_{\bar{\zeta}}}{\zeta - z}\, d\xi\, d\eta$$

$$= T(\eta_{\bar{z}}).$$

Hence

$$\int_\Omega Tf\, \eta_{\bar{z}}\, dx\, dy = -\frac{1}{\pi}\int_\Omega\int_\Omega \frac{f(\zeta)\, \eta_{\bar{z}}(z)}{\zeta - z}\, d\xi\, d\eta\, dx\, dy$$

$$= -\int_\Omega f(\zeta)\, T(\eta_{\bar{\zeta}})\, d\xi\, d\eta$$

$$= -\int_\Omega f\, \eta\, d\xi\, d\eta$$

as required. □

4.3. Regularity for the nonhomogeneous Cauchy–Riemann system

Lemma 4.3.1. *Let* $f \in C^{\mu}(\Omega)$ *for some* μ, $0 < \mu < 1$. *Then the singular integral*

$$h(z) = \Pi f(z) = \Pi_{\Omega} f(z) = -\frac{1}{\pi}\int_{\Omega}\frac{f(\zeta)}{(\zeta - z)^2}\,d\xi\,d\eta \tag{4.17}$$

exists in the sense of the Cauchy principal value at every point $z \in \Omega$. *Furthermore,*

$$h(z) = -\frac{1}{\pi}\int_{\Omega}\frac{f(\zeta)-f(z)}{(\zeta-z)^2}\,d\xi\,d\eta - \frac{f(z)}{\pi}\int_{\Omega\setminus B_{\rho}(z)}\frac{d\xi\,d\eta}{(\zeta-z)^2} \tag{4.18}$$

for all ρ *such that* $B_{\rho}(z) \subset \Omega$. *Here* Ω *is a bounded open set.*

Proof: Let $\Omega_{\epsilon} = \Omega\setminus B_{\epsilon}(z)$, $0 < \epsilon \le \rho$, then

$$\Pi_{\Omega_{\epsilon}} f(z) = -\frac{1}{\pi}\int_{\Omega_{\epsilon}}\frac{f(\zeta)-f(z)}{(\zeta-z)^2}\,d\xi\,d\eta - \frac{f(z)}{\pi}\int_{\Omega_{\epsilon}}\frac{d\xi\,d\eta}{(\zeta-z)^2}.$$

By the Gauß–Green theorem (4.7), it follows that

$$\begin{aligned}
\int_{\Omega_{\epsilon}}\frac{d\xi\,d\eta}{(\zeta-z)^2} &= \int_{\Omega_{\rho}}\frac{d\xi\,d\eta}{(\zeta-z)^2} + \int_{B_{\rho}\setminus B_{\epsilon}}\frac{d\xi\,d\eta}{(\zeta-z)^2}\\
&= \int_{\Omega_{\rho}}\frac{d\xi\,d\eta}{(\zeta-z)^2} - \int_{B_{\rho}\setminus B_{\epsilon}}\frac{\partial}{\partial\zeta}\Big(\frac{1}{\zeta-z}\Big)\,d\xi\,d\eta\\
&= \int_{\Omega_{\rho}}\frac{d\xi\,d\eta}{(\zeta-z)^2} - \frac{1}{2i}\int_{\partial B_{\rho}}\frac{d\bar\zeta}{\zeta-z} + \frac{1}{2i}\int_{\partial B_{\epsilon}}\frac{d\bar\zeta}{\zeta-z}\\
&= \int_{\Omega_{\rho}}\frac{d\xi\,d\eta}{(\zeta-z)^2}.
\end{aligned}$$

This implies the statement. □

Lemma 4.3.2. *Suppose that* Ω *is a smooth domain. Then*

$$h(z) = -\frac{1}{\pi}\int_{\Omega}\frac{f(\zeta)-f(z)}{(\zeta-z)^2}\,d\xi\,d\eta - f(z)\,\Phi'(z), \tag{4.19}$$

where

$$\Phi(z) = \frac{1}{2\pi i}\int_{\partial\Omega}\frac{\bar\zeta\,d\zeta}{\zeta-z}. \tag{4.20}$$

Proof: By Lemma 4.3.1, for all ρ with $B_{\rho}(z) \subset \Omega$,

$$h(z) = -\frac{1}{\pi}\int_{\Omega}\frac{f(\zeta)-f(z)}{(\zeta-z)^2}\,d\xi\,d\eta - \frac{f(z)}{\pi}\int_{\Omega_{\rho}}\frac{d\xi\,d\eta}{(\zeta-z)^2}.$$

Now

$$\frac{1}{\pi}\int_{\Omega_\rho}\frac{d\xi\,d\eta}{(\zeta-z)^2} = -\frac{1}{\pi}\int_{\Omega_\rho}\frac{\partial}{\partial\zeta}\Big(\frac{1}{\zeta-z}\Big)d\xi\,d\eta$$

$$= \frac{1}{2\pi i}\int_{\partial\Omega}\frac{d\bar\zeta}{\zeta-z}$$

$$= \frac{1}{2\pi i}\int_{\partial\Omega}\frac{\bar\zeta\,d\zeta}{(\zeta-z)^2}$$

$$= \frac{1}{2\pi i}\frac{\partial}{\partial z}\int_{\partial\Omega}\frac{\bar\zeta\,d\zeta}{\zeta-z}$$

as required. □

<u>Lemma 4.3.3.</u> *Suppose that* Ω *is a bounded open set,* $\Omega \subset B_R(0)$. *Let* $f \in C^\mu(\Omega)$, $0<\mu<1$, *be such that*

(4.21) $$\sup_\Omega |f| \le a, \quad [f]_\mu^\Omega \le b.$$

Then $h = \Pi f \in C^\mu_{loc}(\Omega)$, *and for all* Ω', $\Omega' \subset\subset \Omega$,

(4.22) $$[h]_\mu^{\Omega'} \le C(\mu, a, b, R, \mathrm{dist}(\Omega', \partial(\Omega)).$$

<u>Proof:</u> Assume w.l.o.g. that $\Omega = B_R(0)$, and let $z_1, z_2 \in B_R$, $z_1 \ne z_2$. Then

$$h(z_1) - h(z_2) = \frac{1}{\pi}\int_{B_R}\Big[\frac{1}{(\zeta-z_2)^2} - \frac{1}{(\zeta-z_1)^2}\Big] f(\zeta)\,d\xi\,d\eta.$$

By partial fractions decomposition:

$$\frac{1}{(\zeta-z_2)^2(\zeta-z_1)} = \frac{1}{z_2-z_1}\frac{1}{(\zeta-z_2)^2} - \frac{1}{(z_2-z_1)^2}\Big[\frac{1}{\zeta-z_2} - \frac{1}{\zeta-z_1}\Big],$$

$$\frac{1}{(\zeta-z_1)^2(\zeta-z_2)} = -\frac{1}{z_2-z_1}\frac{1}{(\zeta-z_1)^2} + \frac{1}{(z_2-z_1)^2}\Big[\frac{1}{\zeta-z_2} - \frac{1}{\zeta-z_1}\Big]$$

and

$$\frac{1}{(\zeta-z_2)^2} - \frac{1}{(\zeta-z_1)^2} = (z_2-z_1)\Big[\frac{1}{(\zeta-z_2)^2(\zeta-z_1)} + \frac{1}{(\zeta-z_1)^2(\zeta-z_2)}\Big].$$

Hence

$$h(z_1) - h(z_2) = \frac{z_2-z_1}{\pi}\Big\{\int_{B_R}\frac{f(\zeta)-f(z_2)}{(\zeta-z_2)^2(\zeta-z_1)}d\xi\,d\eta + \int_{B_R}\frac{f(\zeta)-f(z_1)}{(\zeta-z_1)^2(\zeta-z_2)}d\xi\,d\eta$$
$$+ \int_{B_R}\Big[\frac{f(z_2)}{(\zeta-z_2)^2(\zeta-z_1)} + \frac{f(z_1)}{(\zeta-z_1)^2(\zeta-z_2)}\Big]d\xi\,d\eta\Big\}.$$

We employ partial fractions decomposition, Lemma 4.3.2 for $f \equiv 1$ and Lemma 4.2.1 for $f(z) = \bar{z}$ to obtain

$$(4.23)\quad \int_{B_R} \frac{d\xi\, d\eta}{(\zeta-z_2)^2(\zeta-z_1)} = \frac{1}{z_2-z_1}\int_{B_R} \frac{d\xi d\eta}{(\zeta-z_2)^2} - \frac{1}{(z_2-z_1)^2}\left[\int_{B_R} \frac{d\xi d\eta}{\zeta-z_2} - \int_{B_R}\frac{d\xi d\eta}{\zeta-z_1}\right],$$

$$\frac{1}{\pi}\int_{B_R} \frac{d\xi d\eta}{(\zeta-z_2)^2} = \Phi'(z_2), \qquad \Phi(z) = \frac{1}{2\pi i}\int_{\partial B_R} \frac{\bar{\zeta} d\zeta}{\zeta - z},$$

$$\bar{z}_2 = \frac{1}{2\pi i}\int_{\partial B_R} \frac{\bar{\zeta} d\zeta}{\zeta-z_2} - \frac{1}{\pi}\int_{B_R}\frac{d\xi d\eta}{\zeta-z_2} = \Phi(z_2) - \frac{1}{\pi}\int_{B_R}\frac{d\xi d\eta}{\zeta-z_2}.$$

Therefore

$$\frac{z_2-z_1}{\pi}\int_{B_R}\frac{d\xi\, d\eta}{(\zeta-z_2)^2(\zeta-z_1)} = \Phi'(z_2) - \frac{1}{z_2-z_1}(\bar{z}_1 - \bar{z}_2 + \Phi(z_2) - \Phi(z_1))$$

$$(4.24)\qquad = \frac{\bar{z}_2-\bar{z}_1}{z_2-z_1} + \Phi'(z_2) - \frac{\Phi(z_2)-\Phi(z_1)}{z_2-z_1},$$

$$\frac{z_2-z_1}{\pi}\int_{B_R}\frac{d\xi\, d\eta}{(\zeta-z_1)^2(\zeta-z_2)} = -\Phi'(z_1) + \frac{1}{z_2-z_1}(\bar{z}_1 - \bar{z}_2 + \Phi(z_2) - \Phi(z_1))$$

$$= -\frac{\bar{z}_2-\bar{z}_1}{z_2-z_1} - \Phi'(z_1) + \frac{\Phi(z_2)-\Phi(z_1)}{z_2-z_1},$$

which in turn yields

$$h(z_1) - h(z_2) = \frac{z_2-z_1}{\pi}\left\{\int_{B_R}\frac{f(\zeta)-f(z_2)}{(\zeta-z_2)^2(\zeta-z_1)}d\xi d\eta + \int_{B_R}\frac{f(\zeta)-f(z_1)}{(\zeta-z_1)^2(\zeta-z_2)}d\xi d\eta\right.$$

$$\left. + (f(z_2) - f(z_1))\left\{\frac{\bar{z}_2-\bar{z}_1}{z_2-z_1} + \Phi'(z_2) - \frac{\Phi(z_2)-\Phi(z_1)}{z_2-z_1}\right\}\right\} + f(z_1)(\Phi'(z_2) - \Phi'(z_1)).$$

The first two singular integrals are estimated by Lemma 4.1.1 because $f \in C^{\mu}(B_R)$, and the Cauchy type integral $\Phi(z)$ is analytic. This proves the lemma. □

<u>Remark 4.3.4.</u> The Cauchy type integral

$$(4.25)\qquad \Phi(z) = \frac{1}{2\pi i}\int_{\partial\Omega}\frac{\bar{\zeta}d\zeta}{\zeta-z}$$

belongs to $C^{k,\mu}$ (up to the boundary), if $\partial\Omega \in C^{k,\mu}$. Therefore $h = \Pi f \in C^{\mu}(\Omega)$ together with an estimate of the form

$$(4.26)\qquad [h]_{\mu}^{\Omega} \le C(\mu, a, b, \Omega).$$

Theorem 4.3.5. *Suppose that* Ω *is a bounded open set, and let* $f \in C^{\mu}(\Omega)$ *for some* μ, $0 < \mu < 1$. *Then*

$$g(z) = Tf(z) = -\frac{1}{\pi}\int_{\Omega}\frac{f(\zeta)}{\zeta - z}d\xi\,d\eta \tag{4.27}$$

belongs to $C^{1,\mu}_{loc}(\Omega)$. g *satisfies the Cauchy–Riemann system* (4.10),

$$g_{\bar z} = f, \qquad \textit{and} \qquad g_z = h, \tag{4.28}$$

where

$$h(z) = \Pi f(z) = -\frac{1}{\pi}\int_{\Omega}\frac{f(\zeta)}{(\zeta - z)^2}d\xi\,d\eta. \tag{4.30}$$

Proof: Assume w.l.o.g. that $\Omega = B_R(0)$. For $z, z_0 \in B_R$, $z \neq z_0$, we then have by partial fractions decomposition (see (4.23)) and by (4.24),

$$\begin{aligned}
&\frac{g(z)-g(z_0)}{z-z_0} - h(z_0)\\
&= \frac{1}{z-z_0}\left\{\frac{1}{\pi}\int_{B_R}\frac{f(\zeta)}{\zeta - z_0}d\xi\,d\eta - \frac{1}{\pi}\int_{B_R}\frac{f(\zeta)}{\zeta - z}d\xi\,d\eta\right\} + \frac{1}{\pi}\int_{B_R}\frac{f(\zeta)}{(\zeta - z_0)^2}d\xi\,d\eta\\
&= \frac{z_0-z}{\pi}\int_{B_R}\frac{f(\zeta)-f(z_0)}{(\zeta-z_0)^2(\zeta-z)}d\xi\,d\eta + \frac{z_0-z}{\pi}f(z_0)\int_{B_R}\frac{d\xi\,d\eta}{(\zeta-z_0)^2(\zeta-z)}\\
&= \frac{z_0-z}{\pi}\int_{B_R}\frac{f(\zeta)-f(z_0)}{(\zeta-z_0)^2(\zeta-z)}d\xi\,d\eta + f(z_0)\left[\frac{\bar z_0-\bar z}{z_0-z} + \Phi'(z_0) - \frac{\Phi(z_0)-\Phi(z)}{z_0-z}\right],
\end{aligned}$$

where

$$\Phi(z) = \frac{1}{2\pi i}\int_{\partial B_R}\frac{\bar\zeta d\zeta}{\zeta - z}.$$

Put $z = z_0 + |z-z_0|e^{i\vartheta}$, then by Lemma 4.1.1,

$$\lim_{z\to z_0}\frac{g(z)-g(z_0)}{z-z_0} = h(z_0) + e^{-2i\vartheta}f(z_0).$$

Setting $\vartheta = 0$ resp. $\vartheta = \frac{\pi}{2}$ we obtain

$$\frac{\partial g}{\partial x} = h + f, \qquad \frac{\partial g}{\partial y} = ih - if,$$

from which the statement follows. □

4.4. The similarity principle

Definition 4.4.1. A *pseudoanalytic function* is a complex valued function $w(z)$ of class $C^1_{loc}(\Omega)$ which satisfies a differential inequality of the form

$$|w_{\bar{z}}| \leq M\,|w|. \tag{4.31}$$

Lemma 4.4.2. *Let* $w(z)$ *be a pseudoanalytic function in* Ω. *Assume that*

$$w(z) = e^{g(z)}\,\Phi(z) \tag{4.32}$$

with continuous functions g, Φ *such that* Φ *is analytic on the set* $\Omega_0 = \{z \in \Omega \mid w(z) \neq 0\}$. *Then* Φ *is analytic in* Ω.

Proof: Let $z_0 \in \Omega \backslash \Omega_0$. Because isolated singularities are removable, we can assume that there exists a sequence $\{z_k\}_{k=1}^{\infty}$ of points in $\Omega \backslash \Omega_0$, $z_k \neq z_0$, such that

$$z_k = z_0 + r_k e^{i\vartheta_k} \quad \text{with} \quad r_k \longrightarrow 0,\ \vartheta_k \longrightarrow \vartheta$$

as $k \longrightarrow \infty$. By the mean value theorem,

$$0 = \int_0^1 \{ w_x(z_0 + \tau(z_k - z_0))\,(x_k - x_0) + w_y(\dots)\,(y_k - y_0)\}\,d\tau$$

$$= r_k \int_0^1 \{ w_x(\dots)\cos\vartheta_k + w_y(\dots)\sin\vartheta_k\}\,d\tau.$$

We divide by r_k and let $k \longrightarrow \infty$ to obtain

$$w_x(z_0)\cos\vartheta + w_y(z_0)\sin\vartheta = 0.$$

On the other hand, the inequality (4.31) gives

$$w_x(z_0) + i\,w_y(z_0) = 0.$$

Hence

$$w_x(z_0) = w_y(z_0) = 0,$$

from which $w(z) = o(|z - z_0|)$ and therefore $\Phi(z) = o(|z - z_0|)$ as $z \longrightarrow z_0$. This proves $\Phi'(z_0) = 0$ and hence the analyticity of Φ in Ω. □

Lemma 4.4.3. *Let* $w \in C^1_{loc}(\Omega, \mathbb{C})$ *not vanish in* Ω. *Let* R, $0 < R < 1$, *be so small that*

$$|w(z) - w(z_0)| \le \frac{1}{2} |w(z_0)| \tag{4.33}$$

for $z \in B_R = B_R(z_0)$. *Then the function*

$$\log w(z) = \sum_{k=1}^{\infty} \frac{(-1)^{k-1}}{k} \left[\frac{w(z) - w(z_0)}{w(z_0)}\right]^k \tag{4.34}$$

is of class $C^1(B_R)$ *and satisfies*

$$(\log w)_{\bar{z}} = \frac{w_{\bar{z}}}{w}. \tag{4.35}$$

Proof: For $z \in B_R$,

$$\begin{aligned}
(\log w)_{\bar{z}} &= \sum_{k=1}^{\infty} (-1)^{k-1} \left[\frac{w(z) - w(z_0)}{w(z_0)}\right]^{k-1} \frac{w_{\bar{z}}}{w(z_0)} \\
&= \sum_{k=0}^{\infty} (-1)^k \left[\frac{w(z)}{w(z_0)} - 1\right]^k \frac{w(z)}{w(z_0)} \frac{w_{\bar{z}}}{w(z)} \\
&= \sum_{k=0}^{\infty} (-q)^k (1+q) \frac{w_{\bar{z}}}{w} \\
&= \frac{w_{\bar{z}}}{w},
\end{aligned}$$

where $q = w(z)/w(z_0) - 1$. □

Theorem 4.4.4. *Suppose that* $w(z)$ *is pseudoanalytic with constant* M *in a domain* Ω, $\Omega \subset B_R$. *Then we have the representation*

$$w(z) = e^{g(z)} \Phi(z), \tag{4.36}$$

with an analytic function Φ *in* Ω *and a Hölder continuous function* g, *which satisfies for all* μ, $0 < \mu < 1$, *the estimates*

$$\sup_{B_R} |g| \le 4MR, \tag{4.37}$$

$$[g]_{\mu}^{B_R} \le C(\mu, M, R). \tag{4.38}$$

The representation formula (4.36) is the similarity principle of Bers–Vekua [BS1], [VE1]. The function Φ is called the analytic divisor, and g is the logarithmic difference of the pseudo-analytic function w. The similarity principle holds also for weakly differentiable functions. The following proof is based on [BS2] and [H2].

Proof of Theorem 4.4.4: Let $\Omega_0 = \{z \in \Omega \mid w(z) \neq 0\} \neq \emptyset$ and set

$$g(z) = T_{\Omega_0}\left[\frac{w_{\bar{z}}}{w}\right] = -\frac{1}{\pi}\int_{\Omega_0} \frac{w_{\bar{\zeta}}}{w}\frac{d\xi\, d\eta}{\zeta - z}.$$

By Lemma 4.2.2, g satisfies the estimates

$$\sup_{B_R} |g| \le 4MR, \qquad [g]_\mu^{B_R} \le C(\mu, M, R)$$

for all μ, $0 < \mu < 1$.

Let $B_\rho = B_\rho(z_0) \subset \Omega_0$ so that

$$|w(z) - w(z_0)| \le \tfrac{1}{2}|w(z_0)| \quad \text{for} \quad z \in B_\rho.$$

Then Lemma 4.4.3 implies that

$$(\log w)_{\bar{z}} = \frac{w_{\bar{z}}}{w} = f$$

for $z \in B_\rho$. In order to show that

$$g_{\bar{z}} = f, \tag{4.39}$$

let $\{f^{(k)}(z)\}_{k=1}^{\infty}$ be a sequence of polynomials converging uniformly to f in $\bar{B}_\rho(z_0)$. Consider

$$g_1(z) = T_{B_\rho} f(z) = -\frac{1}{\pi}\int_{B_\rho} \frac{f(\zeta)}{\zeta - z} d\xi\, d\eta,$$

$$g_1^{(k)}(z) = T_{B_\rho} f^{(k)}(z) = -\frac{1}{\pi}\int_{B_\rho} \frac{f^{(k)}(\zeta)}{\zeta - z} d\xi\, d\eta,$$

$$g_2(z) = T_{\Omega_0 \setminus \bar{B}_\rho} f(z) = -\frac{1}{\pi}\int_{\Omega_0 \setminus \bar{B}_\rho} \frac{f(\zeta)}{\zeta - z} d\xi\, d\eta.$$

Then g_2 is analytic in $B_\rho(z_0)$. Furthermore, $g_1^{(k)}(z) \to g_1(z)$ uniformly in $\bar{B}_\rho(z_0)$ and by Theorem 4.3.5, $g_1^{(k)} \in C^{1,\mu}_{loc}(B_\rho)$ and

$$(g_1^{(k)})_{\bar{z}} = f^{(k)}.$$

Hence

$$(g_1^{(k)} - \log w)_{\bar{z}} \longrightarrow 0$$

uniformly in B_ρ. By Lemma 4.2.1,

$$(g_1^{(k)} - \log w)(z) = \frac{1}{2\pi i}\int_{\partial B_\rho} \frac{(g_1^{(k)} - \log w)(\zeta)}{\zeta - z} d\zeta - \frac{1}{\pi}\int_{B_\rho} \frac{(f^{(k)} - f)(\zeta)}{\zeta - z} d\xi\, d\eta,$$

and letting $k \longrightarrow \infty$,

$$(g_1 - \log w)(z) = \frac{1}{2\pi i}\int_{\partial B_\rho} \frac{(g_1 - \log w)(\zeta)}{\zeta - z} d\zeta,$$

i.e., $g_1 - \log w$ is analytic in B_ρ. Therefore $g_1 \in C^1_{loc}(B_\rho)$ and

$$(g_1)_{\bar{z}} = f = \frac{w_{\bar{z}}}{w}.$$

Now $g = g_1 + g_2$, $(g_2)_{\bar{z}} = 0$, hence $g \in C^1_{loc}(B_\rho)$ and (4.39) is therefore true for $z \in \Omega_0$. For $z \in \Omega$ set

$$\Phi(z) = w e^{-g(z)}.$$

Then $\Phi \in C^1_{loc}(\Omega_0) \cap C^0_{loc}(\Omega)$ and we have for $z \in \Omega_0$,

$$\Phi_{\bar{z}} = w_{\bar{z}} e^{-g} - w g_{\bar{z}} e^{-g} = 0.$$

Hence Φ is analytic in Ω_0 and Lemma 4.4.2 implies the statement. □

4.5. Harnack type inequalities for pseudoanalytic functions

Proposition 4.5.1. *Let* $w(z)$ *be pseudoanalytic in* $B_R(0)$ *with constant* M. *Furthermore let*

(4.40) $$\sup_{B_R} |w| \le K,$$

and let $w(z) \neq 0$ *for* $z \in B_R$. *Then, for* $z \in B_\rho(0)$, $0 < \rho < R$, *we have the inequalities*

(4.41) $$|w(z)| \le K^{\frac{2\rho}{R+\rho}} e^{8MR} |w(0)|^{\frac{R-\rho}{R+\rho}},$$

(4.42) $$|w(z)| \ge K^{-\frac{2\rho}{R-\rho}} e^{-\frac{8MR(R+\rho)}{R-\rho}} |w(0)|^{\frac{R+\rho}{R-\rho}}.$$

The Harnack type inequalities (4.41,42) are taken from Heinz [H2] (see Pólya–Szegö [PS] for the case of analytic functions).

Proof of Proposition 4.5.1: The stated inequalities can be written in the form

$$\left|\frac{w(z)}{K}\right| \le e^{8MR}\left|\frac{w(0)}{K}\right|^{\frac{R-\rho}{R+\rho}},$$

$$\left|\frac{w(z)}{K}\right| \ge e^{-\frac{8MR(R+\rho)}{R-\rho}}\left|\frac{w(0)}{K}\right|^{\frac{R+\rho}{R-\rho}}.$$

It therefore suffices to consider the case $K = 1$ only. By virtue of the similarity principle we have the representation

$$w(z) = e^{g(z)}\,\Phi(z),$$

where Φ is analytic in B_R, $\Phi(z) \neq 0$, and

$$|g(z)| \le 4MR.$$

Hence

$$|w(z)| \le e^{4MR}\,|\Phi(z)|,$$

$$|\Phi(z)| \le e^{4MR}\,|w(z)| \le e^{4MR}.$$

The function

$$\varphi(z) = 4MR - \log|\Phi(z)|$$

is therefore nonnegative and harmonic in B_R (compare the upcoming Lemma 4.5.2). We can apply the Harnack inequality for harmonic functions to obtain for $z \in B_\rho$, $0 < \rho < R$, the inequalities

$$\frac{R-\rho}{R+\rho}\varphi(0) \le \varphi(z) \le \frac{R+\rho}{R-\rho}\varphi(0).$$

These can be rewritten in the form

$$\log|\Phi(z)| \le 4MR + \frac{R-\rho}{R+\rho}(\log|\Phi(0)| - 4MR)$$

$$= \frac{8MR\rho}{R+\rho} + \frac{R-\rho}{R+\rho}\log|\Phi(0)|,$$

$$\log|\Phi(z)| \ge -\frac{8MR\rho}{R-\rho} + \frac{R+\rho}{R-\rho}\log|\Phi(0)|,$$

from which

$$|\Phi(z)| \le e^{\frac{8MR\rho}{R+\rho}} |\Phi(0)|^{\frac{R-\rho}{R+\rho}},$$

$$|\Phi(z)| \ge e^{-\frac{8MR\rho}{R-\rho}} |\Phi(0)|^{\frac{R+\rho}{R-\rho}}.$$

Hence

$$\begin{aligned} |w(z)| &\le e^{4MR} |\Phi(z)| \\ &\le e^{\frac{4MR(R+3\rho)}{R+\rho}} |\Phi(0)|^{\frac{R-\rho}{R+\rho}} \\ &\le e^{8MR} |w(0)|^{\frac{R-\rho}{R+\rho}}, \end{aligned}$$

$$|w(z)| \ge e^{-\frac{8MR(R+\rho)}{R-\rho}} |w(0)|^{\frac{R+\rho}{R-\rho}}$$

as required. □

<u>Lemma 4.5.2.</u> *Let* Φ *be an analytic function. Then* $\log|\Phi|$ *is harmonic.*

<u>Proof:</u> *Let* $\Phi = \varphi + i\psi$. Then

$$(\log|\Phi|)_{\bar z} = \frac{1}{|\Phi|^2}(\varphi\varphi_{\bar z} + \psi\psi_{\bar z}),$$

$$\begin{aligned} (\log|\Phi|)_{\bar z z} &= \frac{1}{|\Phi|^2}\{\varphi_z\varphi_{\bar z} + \psi_z\psi_{\bar z} - \frac{2}{|\Phi|^2}(\varphi\varphi_z + \psi\psi_z)(\varphi\varphi_{\bar z} + \psi\psi_{\bar z})\} \\ &= \frac{\psi_{\bar z}}{|\Phi|^2}\{-i\varphi_z + \psi_z - \frac{2}{|\Phi|^2}(\varphi\varphi_z + \psi\psi_z)(-i\varphi + \psi)\} \\ &= \frac{\psi_z\psi_{\bar z}}{|\Phi|^2}\{2 - \frac{2}{|\Phi|^2}(i\varphi + \psi)(-i\varphi + \psi)\} \\ &= 0. \end{aligned}$$

This because $\varphi_{\bar z} = -i\psi_{\bar z}$, $\varphi_z = i\psi_z$. □

Note that the harmonicity of $\log|\Phi|$ is actually an immediate consequence of the definition of the complex logarithm.

Chapter 5. UNIVALENT SOLUTIONS OF BINARY ELLIPTIC SYSTEMS

5.1. Estimates for the gradient from below

Consider a function $w = u + iv \in C^2_{loc}(B)$, $B = B_1(0)$, which solves the binary elliptic system

$$\Delta w = f(z, w(z), Dw(z)). \tag{5.1}$$

Assumption (A5.1). Suppose that $w \in C^2_{loc}(B)$ is a univalent mapping into B with nonvanishing Jacobian

$$Jw = |w_z|^2 - |w_{\bar{z}}|^2 \neq 0, \tag{5.2}$$

and such that

$$\int_B |Dw|^2 dx\,dy \leq M. \tag{5.3}$$

Assumption (A5.2). The function $f(z, w(z), Dw(z))$ is $\mathbb{C}$-valued and continuous on B, satisfying, for some constants a, b,

$$|f(z, w(z), Dw(z))| \leq a\,|Dw(z)|^2 + b\,|Dw(z)| \tag{5.4}$$

for all $z \in B$.

Lemma 5.1.1. *We have for* $z \in B_\rho$, $0 < \rho < 1$, *the inequalities*

$$|Dw| \leq C(a, b, M, \rho)\,|Dw(0)|^{\frac{1-\rho}{1+3\rho}}, \tag{5.5}$$

$$|Dw| \geq c(a, b, M, \rho)\,|Dw(0)|^{\frac{1+3\rho}{1-\rho}}. \tag{5.6}$$

Proof: The results of Chapter 2 are applicable since

$$\begin{aligned} |f| &\leq a\,|Dw|^2 + b\,|Dw| \\ &\leq (a+b)\,|Dw|^2 + b. \end{aligned}$$

Theorem 2.4.4 yields the estimate

$$|Dw| \leq C(a, b, M, R)$$

for $z \in B_R$, $0 < R < 1$, and whence the inequality

$$|\Delta w| \leq C(a, b, M, R)\,|Dw|. \tag{5.7}$$

Trivially,

$$|w_z|^2 = \frac{1}{4}|(u_x - v_y) + i(u_y + v_x)|^2 \le \frac{1}{2}|Dw|^2.$$

On the other hand, we can w.l.o.g. assume that $Jw > 0$ in B by otherwise considering the mapping $\bar{w}$. Hence

$$(5.8) \qquad 0 \le |w_{\bar{z}}|^2 < |w_z|^2,$$

and

$$\begin{aligned} |w_z|^2 &\ge \frac{1}{2}(|w_z|^2 + |w_{\bar{z}}|^2) \\ &= \frac{1}{8}(|(u_x - v_y) + i(u_y + v_x)|^2 \\ &\qquad + |(u_x + v_y) - i(u_y - v_x)|^2) \\ &= \frac{1}{4}|Dw|^2. \end{aligned}$$

The differential inequality

$$(5.9) \qquad |(w_z)_{\bar{z}}| \le C(a,b,M,R)\,|w_z|$$

is therefore satisfied in B_R and $w_z \neq 0$ by (5.8). Proposition 4.5.1 yields for $z \in B_\rho$, $0 < \rho < R$, the inequalities

$$|w_z| \le C(a,b,M,\rho,R)\,|w_z(0)|^{\frac{R-\rho}{R+\rho}},$$

$$|w_z| \ge c(a,b,M,\rho,R)\,|w_z(0)|^{\frac{R+\rho}{R-\rho}}.$$

Now take $R = \frac{1}{2}(1+\rho)$, then

$$\frac{R-\rho}{R+\rho} = \frac{1-\rho}{1+3\rho},$$

from which the statement follows. □

<u>**Theorem 5.1.2.**</u> *Suppose that* w *is a homeomorphism from* $\bar{B}$ *onto* $\bar{B}$ *with* $w(0) = 0$, *which solves the system* (5.1) *and satisfies the assumptions* (A5.1,2). *Then, for* $z \in B_\rho$, $0 < \rho < 1$, *we have the estimate*

$$(5.10) \qquad |Dw| \ge c(a,b,M,\rho) > 0.$$

Proof: By the boundary Courant–Lebesgue lemma, Lemma 1.6.3,

$$|w(z') - w(z'')| \le 4\sqrt{\frac{\pi M}{\log 1/R}}$$

for any $z', z'' \in \bar{B}$ such that $|z' - z''| \le R$, $0 < R < 1/4$. Therefore

$$|w(z)| \ge \tfrac{1}{2}$$

for $|z| = 1 - R = \rho_0$, if R is such that

$$4\sqrt{\frac{\pi M}{\log 1/R}} \le \frac{1}{2}, \qquad 0 < R < \frac{1}{4}.$$

For $|z| = \rho_0$, it follows from $w(0) = 0$ and Lemma 5.1.1, that

$$\begin{aligned}\tfrac{1}{2} &\le |w(z) - w(0)| \\ &= \left|\int_0^1 (w_x(\tau z)\,x + w_y(\tau z)\,y)\,d\tau\right| \\ &\le \int_0^1 |Dw(\tau z)|\,d\tau \\ &\le C(a, b, M, \rho_0)\,|Dw(0)|^{\frac{1-\rho_0}{1+3\rho_0}}.\end{aligned}$$

Using Lemma 5.1.1 again,

$$|Dw| \ge c(a, b, M, \rho) > 0$$

for $z \in B_\rho$, $0 < \rho < 1$, as stated. □

5.2. Estimates for the Jacobian from below

In this section, $w = u + iv \in C^2_{loc}(B)$ is a univalent mapping, satisfying Assumption (A5.1), which solves the Heinz–Lewy system

$$(5.11)\qquad \begin{aligned}\Delta u &= h_1(w)\,|Du|^2 + h_2(w)\,Du \cdot Dv + h_3(w)\,|Dv|^2 + h_4(w)\,Du \wedge Dv, \\ \Delta v &= \tilde{h}_1(w)\,|Du|^2 + \tilde{h}_2(w)\,Du \cdot Dv + \tilde{h}_3(w)\,|Dv|^2 + \tilde{h}_4(w)\,Du \wedge Dv.\end{aligned}$$

Here $Du \wedge Dv = Jw$. To introduce the reader into the subject, we consider only a special, but important case:

<u>Assumption (A5.3).</u> The coefficients $h_1(w(z)),\dots,\tilde h_4(w(z))$ are continuous functions on B with

(5.12) $$\sup_B\{\{h_1|,\dots,|\tilde h_4|\}\le a,$$

and such that the following relations hold:

(5.13)
$$\begin{aligned}\omega_1(w) &= \tilde h_1(w) \equiv 0,\\ \omega_2(w) &= h_1(w)-\tilde h_2(w)\equiv 0,\\ \omega_3(w) &= h_2(w)-\tilde h_3(w)\equiv 0,\\ \omega_4(w) &= h_3(w)\equiv 0.\end{aligned}$$

<u>Remark 5.2.1.</u> The system (5.11) can be written in the form

(5.14)
$$\begin{aligned} u_{z\bar z} &= A(w)\,u_z u_{\bar z} + B(w)\,u_z v_{\bar z} + \overline{B(w)}\,u_{\bar z} v_z + C(w)\,v_z v_{\bar z},\\ v_{z\bar z} &= \tilde A(w)\,u_z u_{\bar z} + \tilde B(w)\,u_z v_{\bar z} + \overline{\tilde B(w)}\,u_{\bar z} v_z + \tilde C(w)\,v_z v_{\bar z},\end{aligned}$$

where

(5.15)
$$\begin{aligned} A &= h_1,\quad B=\tfrac12(h_2-i\,h_4),\quad C=h_4,\\ \tilde A &= \tilde h_1,\quad \tilde B=\tfrac12(\tilde h_2-i\,\tilde h_4),\quad \tilde C=\tilde h_4.\end{aligned}$$

<u>Lemma 5.2.2.</u> *Let* $w\in C^2_{loc}(B_R)$, $0<R\le 1$, *be a solution of the system* (5.11) *with* (A5.3) *satisfied and such that*

(5.16) $$|Dw|\le K.$$

Then, for $\alpha,\beta\in\mathbb{R}$, $\alpha^2+\beta^2=1$, *the real valued function*

(5.17) $$\varphi(z)=\alpha u+\beta v$$

satisfies a differential inequality of the form

(5.18) $$|\Delta\varphi|\le C(a,K)\,|D\varphi|\,.$$

<u>Proof</u>: Using the relations (5.13),

$$\Delta\varphi=\alpha h_1\,|Du|^2+(\alpha h_2+\beta h_1)\,Du\cdot Dv+\beta h_2\,|Dv|^2+(\alpha h_4+\beta\tilde h_4)\,Du\wedge Dv\,.$$

By virtue of $\alpha Du = D\varphi - \beta Dv$,

$$\alpha^2 \Delta\varphi = \alpha h_1 |D\varphi|^2 + (-2\alpha\beta h_1 + \alpha(\alpha h_2 + \beta h_1)) D\varphi \cdot Dv$$
$$+ (\alpha\beta^2 h_1 - \alpha\beta(\alpha h_2 + \beta h_1) + \alpha^2 \beta h_2) |Dv|^2 + \alpha(\alpha h_4 + \beta \tilde{h}_4) D\varphi \wedge Dv$$
$$= \alpha h_1 |D\varphi|^2 + (\alpha^2 h_2 - \alpha\beta h_1) D\varphi \cdot Dv + (\alpha^2 h_4 + \alpha\beta \tilde{h}_4) D\varphi \wedge Dv .$$

Similarly, from $\beta Dv = D\varphi - \alpha Du$,

$$\beta^2 \Delta\varphi = \beta h_2 |D\varphi|^2 + (\beta^2 h_1 - \alpha\beta h_2) D\varphi \cdot Du + (\alpha\beta h_4 + \beta^2 \tilde{h}_4) Du \wedge D\varphi .$$

Therefore

$$\Delta\varphi = (\alpha h_1 + \beta h_2) |D\varphi|^2 + (\beta^2 h_1 - \alpha\beta h_2) D\varphi \cdot Du + (\alpha^2 h_2 - \alpha\beta h_1) D\varphi \cdot Dv$$
$$- (\alpha\beta h_4 + \beta^2 \tilde{h}_4) D\varphi \wedge Du + (\alpha^2 h_4 + \alpha\beta \tilde{h}_4) D\varphi \wedge Dv , \tag{5.19}$$

and the statement follows. □

Lemma 5.2.3. *Let* w *be a solution of* (5.11) *which satisfies the assumption* (A5.1,3). *Then the function* (5.17), $\varphi(z) = \alpha u + \beta v$, $\alpha^2 + \beta^2 = 1$, *satisfies the following inequalities for* $z \in B_\rho$, $0 < \rho < 1$:

$$|D\varphi| \le C(a, M, \rho) \, |D\varphi(0)|^{\frac{1-\rho}{1+3\rho}} , \tag{5.20}$$

$$|D\varphi| \ge c(a, M, \rho) \, |D\varphi(0)|^{\frac{1+3\rho}{1-\rho}} . \tag{5.21}$$

Proof: By Theorem 2.4.4, we have for $z \in B_R$, $0 < R < 1$, the estimate

$$|Dw| \le C(a, M, R) .$$

Hence the function $\varphi(z) = \alpha u + \beta v$ satisfies the differential inequality

$$|\Delta\varphi| \le C(a, M, R) \, |D\varphi|$$

in B_R. This is equivalent to

$$|(\varphi_z)_{\bar{z}}| \le C(a, M, R) \, |\varphi_z| . \tag{5.22}$$

Furthermore

$$\varphi_z \ne 0 \quad \text{in} \quad B , \tag{5.23}$$

because

$$\varphi_z = \frac{1}{2}((\alpha u + \beta v)_x - i(\alpha u + \beta v)_y) = 0$$

iff

$$\alpha u_x + \beta v_x = 0,$$

$$\alpha u_y + \beta v_y = 0,$$

which cannot be true because $Jw \neq 0$ and $(\alpha,\beta) \neq 0$.

Hence we can apply Proposition 4.5.1, which yields for $z \in B_\rho$, $0 < \rho < R$, the inequalities

$$|\varphi_z| \le C(a, M, \rho, R)\, |\varphi_z(0)|^{\frac{R-\rho}{R+\rho}},$$

$$|\varphi_z| \ge c(a, M, \rho, R)\, |\varphi_z(0)|^{\frac{R+\rho}{R-\rho}}.$$

If $R = \frac{1}{2}(1+\rho)$, then

$$\frac{R-\rho}{R+\rho} = \frac{1-\rho}{1+3\rho},$$

and the statement follows. □

<u>Theorem 5.2.4.</u> *Suppose that* w *is a homeomorphism from* $\bar{B}$ *onto* $\bar{B}$ *with* $w(0) = 0$, *which solves the system* (5.11) *and satisfies the assumptions* (A5.1,3). *Then, for* $z \in B_\rho$, $0 < \rho < 1$, *we have an estimate of the form*

(5.24) $$|Jw(z)| \ge c(a, M, \rho) > 0.$$

<u>Proof:</u> Let $(\alpha,\beta) \in \partial B$. Since w maps ∂B onto ∂B, there is an angle ϑ such that

$$w(e^{i\vartheta}) = \alpha + i\beta.$$

Hence for all ρ, $0 < \rho < 1$,

$$|\varphi(\rho e^{i\vartheta}) - 1| = |\alpha(u(\rho e^{i\vartheta}) - \alpha) + \beta(v(\rho e^{i\vartheta}) - \beta)|$$
$$\le |w(\rho e^{i\vartheta}) - w(e^{i\vartheta})|.$$

By the boundary Courant–Lebesgue lemma, Lemma 1.6.3, if we choose R so that

$$4\sqrt{\frac{\pi M}{\log 1/R}} \le \frac{1}{2}, \qquad 0 < R < \frac{1}{4},$$

then, for $\rho_0 = 1-R$,

$$|w(\rho_0 e^{i\vartheta}) - w(e^{i\vartheta})| \le \tfrac{1}{2}.$$

Therefore

$$\begin{aligned} |\varphi(\rho_0 e^{i\vartheta})| &\ge 1 - |\varphi(\rho_0 e^{i\vartheta}) - 1| \\ &\ge 1 - |w(\rho_0 e^{i\vartheta}) - w(e^{i\vartheta})| \\ &\ge \tfrac{1}{2}. \end{aligned}$$

On the other hand, from $\varphi(0) = 0$ and from Lemma 5.2.3, it follows that

$$\begin{aligned} \tfrac{1}{2} &\le |\varphi(\rho_0 e^{i\vartheta}) - \varphi(0)| \\ &\le \int_0^1 |D\varphi(\tau\rho_0 e^{i\vartheta})|\, d\tau \\ &\le C(a, M, \rho_0)\, |D\varphi(0)|^{\frac{1-\rho_0}{1+3\rho_0}}. \end{aligned}$$

Therefore, again by Lemma 5.2.3,

$$|D\varphi| \ge c(a, M, \rho) > 0$$

for $z \in B_\rho$, $0 < \rho < 1$, and for all $\alpha, \beta \in \mathbb{R}$, $\alpha^2 + \beta^2 = 1$, i.e.,

$$(\alpha u_x + \beta v_x)^2 + (\alpha u_y + \beta v_y)^2 \ge c(\alpha^2 + \beta^2) \tag{5.25}$$

for all $\alpha, \beta \in \mathbb{R}$. This implies the stated estimate (5.24).

Consider e. g. the system

$$\begin{aligned} \alpha u_x + \beta v_x &= \xi \\ \alpha u_y + \beta v_y &= \eta. \end{aligned} \tag{5.26}$$

Then by Cramer's rule,

$$(\alpha^2 + \beta^2)(Jw)^2 = \det\begin{bmatrix} \xi & v_x \\ \eta & v_y \end{bmatrix}^2 + \det\begin{bmatrix} u_x & \xi \\ u_y & \eta \end{bmatrix}^2.$$

Now solve (5.26) for $(\xi, \eta) = (1, 0)$ to get

$$(\alpha^2+\beta^2)(Jw)^2 = (v_y^2 + u_y^2)(\xi^2 + \eta^2)$$
$$\geq c(v_y^2 + u_y^2)(\alpha^2 + \beta^2)$$

by (5.25). Therefore

$$(Jw)^2 \geq c(a, M, \rho)(u_y^2 + v_y^2). \tag{5.27}$$

The estimate (5.24) follows by combining (5.27) with

$$(Jw)^2 \geq c(u_x^2 + v_x^2),$$

$$u_x^2 + u_y^2 \geq c. \quad \square$$

Chapter 6. CONFORMAL MAPPINGS WITH RESPECT TO A RIEMANNIAN METRIC

6.1. The Cauchy–Riemann–Beltrami system

Consider a Riemannian metric

$$ds^2 = a(x,y)\,dx^2 + 2b(x,y)\,dx\,dy + c(x,y)\,dy^2, \tag{6.1}$$

$$\Delta = ac - b^2 > 0,$$

in a simply connected domain Ω. Here x and y are coordinates in a neighborhood $\mathcal{N}$ of $(x_0, y_0) \in \Omega$ and $a, b, c \in C^0(\mathcal{N})$. We wish to reduce ds^2 to the canonical form

$$ds^2 = \Lambda\,(du^2 + dv^2), \qquad \Lambda \neq 0, \tag{6.2}$$

in the small and later in the large. This means that we wish to introduce new parameters $u(x,y)$ and $v(x,y)$ in a neighborhood of (x_0, y_0) such that

$$Jw = u_x v_y - u_y v_x \neq 0 \tag{6.3}$$

and

$$\begin{aligned} a\,dx^2 + 2b\,dx\,dy + c\,dy^2 &= \Lambda\,((u_x\,dx + u_y\,dy)^2 + (v_x\,dx + v_y\,dy)^2) \\ &= \Lambda\,((u_x^2 + v_x^2)\,dx^2 + 2\,(u_x u_y + v_x v_y)\,dx\,dy + (u_y^2 + v_y^2)\,dy^2), \end{aligned}$$

i.e., with $\lambda = 1/\Lambda$,

$$u_x^2 + v_x^2 = \lambda a, \tag{6.4}$$

$$u_x u_y + v_x v_y = \lambda b, \tag{6.5}$$

$$u_y^2 + v_y^2 = \lambda c. \tag{6.6}$$

Furthermore,

$$\begin{aligned} \lambda^2 \Delta &= (u_x^2 + v_x^2)(u_y^2 + v_y^2) - (u_x u_y + v_x v_y)^2 \\ &= (Jw)^2. \end{aligned}$$

Hence

$$\lambda = \frac{Jw}{\sqrt{\Delta}} \tag{6.7}$$

if

$$a\,Jw \geq 0. \tag{6.8}$$

The equations (6.4, 5, 6) are the *conformality relations*, meaning that $w=(u,v)$ maps a neighborhood of (x_0,y_0) comformally with respect to ds^2 onto a neighborhood of $(u_0,v_0)=w(x_0,y_0)$. This if (6.8) is satisfied, otherwise the mapping would be anti–conformal, because $\bar{w}=(u,-v)=u-iv$ would satisfy the relations (6.4, 5, 6) and (6.8). The variables u and v are called (locally) *uniformizing parameters.*

By multiplying (6.4) with v_y and (6.5) with v_x, one obtains by subtracting the resulting equations:

$$u_x(u_x v_y - u_y v_x) = \lambda(a v_y - b v_x),$$

and similarly,

$$u_y(u_x v_y - u_y v_x) = \lambda(b v_y - c v_x).$$

Hence:

<u>Remark 6.1.1.</u> If (6.8) is satisfied, then the conformality relations (6.4, 5, 6) imply the *Cauchy–Riemann–Beltrami system*

(6.9) $$\sqrt{\Delta}\, u_x = -b v_x + a v_y,$$

(6.10) $$\sqrt{\Delta}\, u_y = -c v_x + b v_y.$$

This system can be written in the equivalent form

(6.11) $$\sqrt{\Delta}\, v_x = b u_x - a u_y$$

(6.12) $$\sqrt{\Delta}\, v_y = c u_x - b u_y.$$

<u>Remark 6.1.2.</u> The functions $u=u(x,y)$, $v=v(x,y)$ satisfy the integrability conditions

(6.13) $$\frac{\partial}{\partial x}\left[\frac{c u_x - b u_y}{\sqrt{\Delta}}\right] + \frac{\partial}{\partial y}\left[\frac{-b u_x + a u_y}{\sqrt{\Delta}}\right] = 0,$$

(6.14) $$\frac{\partial}{\partial x}\left[\frac{c v_x - b v_y}{\sqrt{\Delta}}\right] + \frac{\partial}{\partial y}\left[\frac{-b v_x + a v_y}{\sqrt{\Delta}}\right] = 0.$$

<u>Remark 6.1.3.</u> In the case $a=c$, $b=0$, the conformality relations (6.4, 5, 6) become

(6.15) $$u_x^2 + v_x^2 = u_y^2 + v_y^2,$$

(6.16) $$u_x u_y + v_x v_y = 0,$$

and the Beltrami system (6.8, 9) reduces to the Cauchy–Riemann system

(6.17)
$$u_x = v_y, \quad u_y = -v_x,$$

i.e.,

(6.18)
$$w_{\bar{z}} = 0.$$

Remark 6.1.4. Suppose that $\tilde{u}(x,y)$ and $\tilde{v}(x,y)$ are also locally uniformizing, i.e.,

(6.19)
$$\begin{aligned} ds^2 &= \Lambda\,(du^2 + dv^2) \\ &= \tilde{\Lambda}\,(d\tilde{u}^2 + d\tilde{v}^2), \end{aligned}$$

and

(6.20)
$$a\,J\tilde{w} > 0.$$

Then

(6.21)
$$\tilde{w}(u + i\,v) = \tilde{u} + i\,\tilde{v}$$

satisfies the Cauchy–Riemann system

(6.22)
$$\frac{\partial \tilde{u}}{\partial u} = \frac{\partial \tilde{v}}{\partial v}, \quad \frac{\partial \tilde{u}}{\partial v} = -\frac{\partial \tilde{v}}{\partial u},$$

i.e.,

(6.23)
$$\tilde{w}_{\bar{w}} = 0$$

and

(6.24)
$$J\tilde{w} = |\tilde{w}_w|^2 \neq 0.$$

Coordinate changes are therefore conformal mappings. More generally, if uniformizers only satisfy $Jw \neq 0$, then coordinate changes are either conformal or anti-conformal. Conversely, any analytic function $\Phi(w)$ with $\Phi' \neq 0$ defines uniformizing coordinates.

Remark 6.1.5. Suppose that $w = (u, v)$ solves the Beltrami system (6.9, 10), and assume only that

(6.25)
$$v_x^2 + v_y^2 > 0.$$

Then

(6.26)
$$\begin{aligned} \sqrt{\Delta}\,Jw &= \sqrt{\Delta}\,(u_x v_y - u_y v_x) \\ &= (-b\,v_x + a\,v_y)\,v_y - (-c\,v_x + b\,v_y)\,v_x \\ &= a\,v_y^2 - 2\,b\,v_x v_y + c\,v_x^2. \end{aligned}$$

Hence

$$\text{(6.27)} \qquad a\,Jw > 0 .$$

Put

$$\text{(6.28)} \qquad \lambda = \frac{Jw}{\sqrt{\Delta}} .$$

Then, from (6.9, 11),

$$\lambda a = \frac{a}{\sqrt{\Delta}}(u_x v_y - u_y v_x)$$

$$= \frac{a}{\sqrt{\Delta}}\left[u_x \frac{\sqrt{\Delta}u_x + b v_x}{a} - \frac{-\sqrt{\Delta}v_x + b u_x}{a} v_x\right]$$

$$\text{(6.29)} \qquad = u_x^2 + v_x^2$$

etc., i.e., the conformality relations can be derived from the Beltrami system, and they are therefore equivalent to it.

<u>Remark 6.1.6.</u> Note that the Beltrami system for the inverse mapping (x, y) reads as

$$\text{(6.30)} \qquad \sqrt{\Delta}x_u = b x_v + c y_v ,$$

$$\text{(6.31)} \qquad \sqrt{\Delta}x_v = -b x_u - c y_u ,$$

resp.,

$$\text{(6.32)} \qquad \sqrt{\Delta}y_u = -a x_v - b y_v ,$$

$$\text{(6.33)} \qquad \sqrt{\Delta}y_v = a x_u + b y_u .$$

The equivalent conformality relations are

$$\text{(6.34)} \qquad x_u^2 + x_v^2 = \frac{\lambda c}{(Jw)^2} ,$$

$$\text{(6.35)} \qquad x_u y_u + x_v y_v = -\frac{\lambda b}{(Jw)^2} ,$$

$$\text{(6.36)} \qquad y_u^2 + y_v^2 = \frac{\lambda a}{(Jw)^2} ,$$

resp.,

$$\frac{a}{\sqrt{\Delta}} = \frac{y_u^2 + y_v^2}{x_u y_v - x_v y_u}, \tag{6.37}$$

$$-\frac{b}{\sqrt{\Delta}} = \frac{x_u y_u + x_v y_v}{x_u y_v - x_v y_u}, \tag{6.38}$$

$$\frac{c}{\sqrt{\Delta}} = \frac{x_u^2 + x_v^2}{x_u y_v - x_v y_u}. \tag{6.39}$$

Remark 6.1.7. If the metric ds^2 is of class C^1, and if the mapping $(x,y) = (x(u,v), y(u,v))$ belongs to C^2, then, by differentiating the systems (6.30, 31) and (6.32, 33), one obtains the following differential equations:

$$\begin{aligned} L_\Delta x &= (\sqrt{\Delta}\, x_u)_u + (\sqrt{\Delta}\, x_v)_v \\ &= b_u x_v + c_u y_v - b_v x_u - c_v y_u \\ &= (c_x - b_y)(x_u y_v - x_v y_u), \end{aligned} \tag{6.40}$$

$$L_\Delta y = (a_y - b_x)(x_u y_v - x_v y_u). \tag{6.41}$$

The system (6.38, 39) remains true in the weak sense if $z = (x,y)$ is only of class C^1.

Remark 6.1.8. Suppose that

$$ds^2 = (A(x,y) + z_{xx})\,dx^2 + 2\,(B(x,y) + z_{xy})\,dx\,dy + (C(x,y) + z_{yy})\,dy^2 \tag{6.42}$$

with a function z of class C^3. Then the system (6.40, 41) takes the form

$$L_\Delta x = (C_x - B_y)(x_u y_v - x_v y_u), \tag{6.40$'$}$$

$$L_\Delta y = (A_y - B_x)(x_u y_v - x_v y_u). \tag{6.41$'$}$$

6.2. Existence of locally uniformizing parameters

Lemma 6.2.1. *Suppose that* a, b *and* c *are real analytic functions in* Ω *which satisfy*

$$\Delta = ac - b^2 > 0. \tag{6.43}$$

Let $(x_0, y_0) \in \Omega$. *Then there is a real analytic univalent mapping* $w = (u,v)$ *in some neighborhood of* (x_0, y_0), *which solves the Beltrami system*

(6.44) $$\sqrt{\Delta}\,u_x = -b\,v_x + a\,v_y,$$

(6.45) $$\sqrt{\Delta}\,u_y = -c\,v_x + b\,v_y,$$

and such that

(6.46) $$a\,Jw = a(u_x v_y - u_y v_x) > 0.$$

Proof: The integrability condition (6.14) can be solved subject to the initial conditions

(6.47) $$v(x_0, y) = 0, \qquad v_x(x_0, y) = 1$$

by the Cauchy–Kovalevsky theorem. Then put

$$u(x,y) = \int_{(x_0,y_0)}^{(x,y)} \left[\frac{-b\,v_x + a\,v_y}{\sqrt{\Delta}}dx - \frac{c\,v_x + b\,v_y}{\sqrt{\Delta}}dy\right].$$

Here the integration is along the straight line segment from (x_0, y_0) to (x, y). The equations (6.44, 45) are then satisfied and (6.46) follows from (6.47) and Remark 6.1.5. □

6.3. Uniformization and second order elliptic systems

Theorem 6.3.1. *Suppose that* ds^2 *is a real analytic Riemannian metric in* Ω, *which in local coordinates* x *and* y *takes the form*

(6.48) $$ds^2 = a(x,y)\,dx^2 + 2\,b(x,y)\,dx\,dy + c(x,y)\,dy^2,$$

$$\Delta = a\,c - b^2 > 0.$$

Furthermore let $\bar{B}_R = \bar{B}_R(x_0, y_0) \subset \Omega$. *Then there is a real analytic homeomorphism* $w = (u, v)$ *from* B_R *onto* $B = \{u^2 + v^2 < 1\}$ *with* $w(x_0, y_0) = 0$ *such that*

(6.49) $$ds^2 = \frac{\sqrt{\Delta}}{Jw}(du^2 + dv^2)$$

and

(6.50) $$a\,Jw = a(u_x v_y - u_y v_x) > 0.$$

Proof: By Lemma 6.2.1 there exist locally uniformizing parameters $w = (u, v)$ for each $(x_0, y_0) \in \Omega$ such that $a\,J > 0$. In particular, by Remark 6.1.4, coordinate changes $\tilde{w}(w)$ are conformal,

$$\frac{\partial\tilde{w}}{\partial\bar{w}} = 0, \qquad \text{and} \qquad \frac{\partial\tilde{w}}{\partial w} \neq 0.$$

Suppose that $(x_0, y_0) \in \partial B_R$. By applying a conformal map Φ as in Remark 6.1.4, there exists a neighborhood $\mathcal{N}(x_0, y_0)$ and a boundary uniformizer which maps $B_R \cap \mathcal{N}(x_0, y_0)$ onto a domain in the upper half plane $v > 0$ so that $\partial B_R \cap \mathcal{N}(x_0, y_0)$ is mapped into a segment of the real axis $v = 0$.

Together with this structure, B_R is therefore a finite simply connected Riemann surface (compare [SHS], p. 25). By the uniformization theorem for finite Riemann surfaces [SHS], p. 61], it follows that B_R can be mapped conformally onto $B = \{u^2 + v^2 < 1\}$ by a real analytic uniformizer $w = (u, v)$ with $w(x_0, y_0) = 0$ and $a \cdot Jw > 0$ as required. □

<u>Theorem 6.3.2.</u> *Let* A, B *and* C *be functions of class* $C^1(\Omega)$ *and let* $z \in C^{2,\mu}(\Omega)$ *for some* μ, $0 < \mu < 1$, *such that*

$$\Delta = (A + z_{xx})(C + z_{yy}) - (B + z_{xy})^2 > 0. \tag{6.51}$$

Let $\bar{B}_R = \bar{B}_R(x_0, y_0) \subset \Omega$. *Then there exists a homeomorphism* $(x, y) = (x(u, v), y(u, v))$ *from* $\bar{B} = \{u^2 + v^2 \leq 1\}$ *onto* $\bar{B}_R$ *of class* $C^{1,\mu}_{loc}(B)$ *with* $x(0) = x_0$, $y(0) = y_0$, *which satisfies the system*

$$L_\Delta x = (C_x - B_y)(x_u y_v - x_v y_u), \tag{6.52}$$

$$L_\Delta y = (A_y - B_x)(x_u y_v - x_v y_u). \tag{6.53}$$

Here

$$L_\Delta = \frac{\partial}{\partial u}\left[\sqrt{\Delta}\frac{\partial}{\partial u}\right] + \frac{\partial}{\partial v}\left[\sqrt{\Delta}\frac{\partial}{\partial v}\right]. \tag{6.54}$$

Furthermore

$$J(x, y) = x_u y_v - x_v y_u \neq 0, \tag{6.55}$$

and

$$\frac{A + z_{xx}}{\sqrt{\Delta}} = \frac{y_u^2 + y_v^2}{J(x, y)}, \tag{6.56}$$

$$-\frac{B + z_{xy}}{\sqrt{\Delta}} = \frac{x_u y_u + x_v y_v}{J(x, y)}, \tag{6.57}$$

$$\frac{C + z_{yy}}{\sqrt{\Delta}} = \frac{x_u^2 + x_v^2}{J(x, y)}. \tag{6.58}$$

Proof: According to Theorem 6.3.1 and Remarks 6.1.6, 7, 8, the statement is true if a, b, c and z are real analytic. This by considering the differential form

$$ds^2 = (A + z_{xx})\,dx^2 + 2(B + z_{xy})\,dx\,dy + (C + z_{yy})\,dy^2$$

in Ω. By the Weierstraß approximation theorem, there exist sequences of polynomials $\{A^{(n)}\}$, $\{B^{(n)}\}$, $\{C^{(n)}\}$, $\{z^{(n)}\}$ such that

$$A^{(n)} \to A, \dots, z^{(n)} \to z,$$
$$A_x^{(n)} \to A_x, \dots, z_{yy}^{(n)} \to z_{yy}$$

uniformly in $\bar{B}_R$. Hence

$$|A^{(n)}|, \dots, |z_{yy}^{(n)}| \le C,$$

and, w.l.o.g.,

$$\Delta^{(n)} \ge c > 0,$$

and

$$[\Delta^{(n)}]_\mu^{B_R} \le C$$

by inspecting the proof of the Weierstraß approximation theorem. From (6.56, 57, 58),

$$\frac{A^{(n)} + z_{xx}^{(n)} + C^{(n)} + z_{yy}^{(n)}}{\sqrt{\Delta^{(n)}}} = \frac{(x_u^{(n)})^2 + (x_v^{(n)})^2 + (y_u^{(n)})^2 + (y_v^{(n)})^2}{J(x^{(n)}, y^{(n)})}$$

in B. Hence

(6.59) $$(x_2^{(n)})^2 + (x_v^{(n)})^2 + (y^{(n)})^2 + (y_v^{(n)})^2 \le C\,J(x^{(n)}, y^{(n)}),$$

and therefore

$$\begin{aligned}\int_B ((x_u^{(n)})^2 + \dots + (y_v^{(n)})^2)\,du\,dv &\le C \int_B J(x,y)\,du\,dv \\ &= C \int_{B_R} dx\,dy \\ &= C R^2.\end{aligned}$$

The regularity theory, Theorem 2.4.4, yields for any ρ, $0<\rho<1$,

$$|x_u^{(n)}|,\dots,|y_v^{(n)}| \le C(R,\rho) \quad \text{in} \quad B_\rho, \tag{6.60}$$

$$[x_u^{(n)}]_\mu^{B_\rho},\dots,[y_v^{(n)}]_\mu^{B_\rho} \le C(R,\rho). \tag{6.61}$$

By passing to a subsequence, there exists functions $x, y \in C^{1,\mu}_{loc}(B)\cap C^0(\bar{B})$ such that, w. l. o. g.,

$$x^{(n)} \longrightarrow x, \quad y^{(n)} \longrightarrow y,$$
$$x_u^{(n)} \longrightarrow x_u, \dots, y_v^{(n)} \longrightarrow y_v$$

uniformly in B_ρ, and (x,y) satisfies the system (6.52, 53) in the weak sense. The mapping $(x,y)=(x(u,v),y(u,v))$ is univalent because the inverses $w^{(n)}=(u^{(n)}(x,y),v^{(n)}(x,y))$ are equicontinuous in $\bar{B}_R$. This is true because the conformality relations (6.4, 6) imply

$$\frac{A^{(n)}+z_{xx}^{(n)}+C^{(n)}+z_{yy}^{(n)}}{\sqrt{\Delta^{(n)}}} = \frac{|Dw^{(n)}|^2}{J(w^{(n)})},$$

and therefore

$$\begin{aligned}\int_{B_R} |Dw^{(n)}|^2\,dx\,dy &\le C\int_{B_R} J(w^{(n)})\,dx\,dy\\ &= C\int_B du\,dv\\ &= C,\end{aligned}$$

from which the equicontinuity follows by the Courant–Lebesgue lemma, Lemma 1.6.3.

In order to conclude that $(x,y)=(x(u,v),y(u,v))$ is a diffeomorphism from B onto B_R, consider the integrability conditions for the inverses $w^{(n)}=(u^{(n)},v^{(n)})$, the elliptic system (6.13, 14):

$$\frac{\partial}{\partial x}\left[\frac{C^{(n)}w_x^{(n)} - B^{(n)}w_y^{(n)}}{\sqrt{\Delta^{(n)}}}\right] + \frac{\partial}{\partial y}\left[\frac{-B^{(n)}w_x^{(n)} + A^{(n)}w_y^{(n)}}{\sqrt{\Delta^{(n)}}}\right] = 0. \tag{6.62}$$

The regularity theory for linear equations, Theorem 2.3.3, gives estimates of the form (6.60, 61) for $w^{(n)}$ and therefore a limit mapping $w=(u,v)$ of class $C^{1,\mu}_{loc}(B_R)\cap C^0(\bar{B}_R)$, which is the inverse of (x,y). This in turn implies the nonvanishing of $J(x,y)$ and the relations (6.56, 57, 58) are therefore satisfied. □

6.4. Metrics with Hölder continuous and measurable coefficients

Theorem 6.3.2 is tailor made for application to Monge–Ampère equations. Let us however note:

Remark 6.4.1. If ds^2 belongs to the Hölder class C^{μ} for some μ, $0 < \mu < 1$, then there exists a uniformizer $w = (u, v)$ which is a diffeomorphism of class $C^{1,\mu}_{loc}$ satisfying the Beltrami system (6.9, 10) and hence the conformality relations (6.4, 5, 6), which can be written in the (usual) form

$$c u_x^2 - 2 b u_x u_y + a u_y^2 = c v_x^2 - 2 b v_x v_y + a v_y^2, \tag{6.63}$$

$$c u_x v_x - b (u_x v_y + u_y v_x) + a u_y v_y = 0. \tag{6.64}$$

A proof of this statement goes e. g. along the lines of the proof of Theorem 6.3.2, except that, instead of employing the system (6.52, 53) for the inverse mapping $(x(u, v), y(u, v))$, one considers the Beltrami system (6.30, 31), resp. (6.32, 33), written in the form

$$a x_u = -b y_u + \sqrt{\Delta} y_v, \tag{6.65}$$

$$a x_v = -\sqrt{\Delta} y_u - b y_v, \tag{6.66}$$

and

$$c y_u = -b x_u - \sqrt{\Delta} x_v, \tag{6.67}$$

$$c y_v = \sqrt{\Delta} x_u - b x_v. \tag{6.68}$$

The integrability conditions are

$$\frac{\partial}{\partial u}\left[\frac{\sqrt{\Delta} y_u - b y_v}{a}\right] + \frac{\partial}{\partial v}\left[\frac{-b y_u + \sqrt{\Delta} y_v}{a}\right] = 0, \tag{6.69}$$

$$\frac{\partial}{\partial u}\left[\frac{\sqrt{\Delta} x_u - b x_v}{c}\right] + \frac{\partial}{\partial v}\left[\frac{b x_u + \sqrt{\Delta} x_v}{c}\right] = 0. \tag{6.70}$$

Note that the coefficients of this system are not symmetric, however, this is no obstacle.

If the coefficients a, b, c are only of class L^{∞}, then the regularity theory of Chapter 2 is not directly applicable anyway. The regularity theory for equations with measurable coefficients yields then the Hölder continuity of the mapping (x, y). The approximation of a metric with measurable coefficients almost everywhere by regular metrics yields therefore via the weak compactness of bounded sets in $W^{1,2}$ the following theorem:

<u>Remark 6.4.2.</u> If ds^2 has measurable, essentially bounded coefficients, then there exists a uniformizer $w = (u, v)$ which is a homeomorphism of class $C^\mu \cap W^{1,2}$ for some μ, $0 < \mu < 1$, satisfying the conformality relations (6.63, 64) almost everywhere.

Chapter 7. LOCAL BEHAVIOR OF SOLUTIONS OF DIFFERENTIAL INEQUALITIES

This chapter develops further the function theory of elliptic equations. Some results are presented about the local behavior of solutions of second order differential inequalities which cannot be deduced directly from the similarity principle.

7.1. The Carleman–Hartman–Wintner theorem

Let $\varphi(z) \in C^1(\Omega)$ be a real-valued function satisfying the differential inequality

$$|\Delta\varphi| \leq C(|D\varphi| + |\varphi|), \tag{7.1}$$

i.e.,

$$-\Delta\varphi = W, \qquad |W(z)| \leq C(|D\varphi(z)| + |\varphi(z)|) \tag{7.2}$$

$(z \in \Omega)$ in the weak sense. Suppose that Ω contains the origin.

<u>Theorem 7.1.1.</u> *Assume that* $\varphi(z) = o(|z|^n)$ *as* $|z| \to 0$ *for some* $n \in \mathbb{N}_0$. *Then*

$$\lim_{\substack{|z|\to 0\\ z\neq 0}} \frac{\varphi_z(z)}{z^n} \tag{7.3}$$

exists.

This theorem is due to Hartman–Wintner [HW 2] and generalizes a result by Carleman [CM]. The following proof is along the lines of [HW 2] and relies on Carleman's integral estimates:

<u>Proof of Theorem 7.1.1:</u> The statement is clearly true for $n = 0$, hence let $n \geq 1$. For convenience assume that $\varphi \in C^2$. Consider the differential equation

$$-4(\varphi_z)_{\bar{z}} = W$$

in a disc $B_R = B_R(0) \subset \Omega$. Multiply it by

$$\eta(z) = z^{-k}(z-\zeta)^{-1}$$

$(k = 1, 2, \ldots, n)$ and integrate over $\Omega_\epsilon = B_R \backslash (B_\epsilon \cup B_\epsilon(\zeta))$ to obtain, by the analyticity of η and by the Gauß–Green theorem, formula (4.6),

$$-4\int_{\Omega_\epsilon} (\varphi_z \eta)_{\bar{z}}\, dx\, dy = 2i \int_{\partial\Omega_\epsilon} \varphi_z \eta\, dz.$$

Hence

$$2i\int_{\partial\Omega_\epsilon} \varphi_z \eta \, dz = \int_{\Omega_\epsilon} W \eta \, dx\, dy .$$

If

(7.4) $$\varphi_z = o(|z|^{k-1}) \quad \text{as} \quad |z| \longrightarrow 0,$$

then

$$\varphi_z h = (\varphi_z z^{-k+1}) z^{-1} (z-\zeta)^{-1} .$$

One can let $\epsilon \longrightarrow 0$ to obtain

$$4\pi \varphi_z(\zeta)\, \zeta^{-k} = -(2\pi i)(2i)\, \varphi_z(\zeta)\, \zeta^{-k}$$

(7.5) $$= -2i\int_{\partial B_R} \varphi_z z^{-k} (z-\zeta)^{-1} dz + \int_{B_R} W z^{-k} (z-\zeta)^{-1} dx\, dy ,$$

and the double integral is absolutely convergent.

It will be shown by induction over k, $1 \le k \le n$, that (7.4) holds. First, (7.4) is true for $k = 1$ because in this case, $\varphi(z) = o(|z|)$, and therefore

$$\varphi_z(0) = \lim_{\substack{z \to 0 \\ z \neq 0}} \frac{\varphi(z) - \varphi(0)}{z - 0} = 0 .$$

Suppose now that (7.4) holds for a k, $1 \le k < n$. Since the double integral in (7.5) is convergent, there exists a constant C such that

(7.6) $$|D\varphi(\zeta)|\, |\zeta|^{-k} \le C\Big\{\int_{B_R} |D\varphi|\, |z|^{-k} |z-\zeta|^{-1} ds + \int_{\partial B_R} (|D\varphi| + |\varphi|)\, |z|^{-k} |z-\zeta|^{-1} dx\, dy\Big\}.$$

This inequality is multiplied by $|\zeta - z_0|^{-1}$, $z_0 \in B_R$, and integrated with respect to $\zeta = (\xi, \eta)$ over B_R to yield:

$$\int_{B_R} |D\varphi(\zeta)|\, |\zeta|^{-k} |\zeta - z_0|^{-1} d\xi\, d\eta \le C\Big\{\int_{\partial B_R} |D\varphi(z)|\, |z|^{-k} \int_{B_R} |z-\zeta|^{-1} |\zeta - z_0|^{-1} d\xi\, d\eta\, ds + \int_{B_R} (|D\varphi(z)| + |\varphi(z)|)\, |z|^{-k} \int_{B_R} |z-\zeta|^{-1} |\zeta - z_0|^{-1} d\xi\, d\eta\, dx\, dy\Big\}.$$

Now use the identity

$$|(z-\zeta)(\zeta-z_0)|^{-1} = |z-z_0|^{-1}\,|(z-\zeta)^{-1}+(\zeta-z_0)^{-1}|,$$

which is easily shown by partial fractions decomposition, and the estimate

$$\int_{B_R} |z-\zeta|^{-1}\,dx\,dy \le C\,R$$

for $\zeta \in B_R$. Then, after relabelling,

$$\int_{B_R} |D\varphi(\zeta)|\,|\zeta|^{-k}\,|z-\zeta|^{-1}\,dx\,dy$$

$$\le C\,R\left\{\int_{\partial B_R} |D\varphi(z)|\,|z|^{-k}|z-\zeta|^{-1}\,ds + \int_{B_R} (|D\varphi(z)|+|\varphi(z)|)\,|z|^{-k}\,|z-\zeta|^{-1}\,dx\,dy\right\}.$$

If $R \le 1/2C$ then

(7.7) $$\int_{B_R} |D\varphi|\,|z|^{-k}\,|z-\zeta|^{-1}\,dx\,dy$$

$$\le C\,R\left\{\int_{\partial B_R} |D\varphi|\,|z|^{-k}\,|z-\zeta|^{-1}\,ds + \int_{B_R} |\varphi|\,|z|^{-k}\,|z-\zeta|^{-1}\,dx\,dy\right\}.$$

Now

$$\varphi(z)\,z^{-k} = O(1) \quad \text{as} \quad |z| \longrightarrow 0.$$

Hence the last integral in (7.7) is $O(1)$ as $\zeta \longrightarrow 0$, and the integral on the l.h.s. of (7.7) is therefore also $O(1)$ as $\zeta \longrightarrow 0$. From (7.6) it follows that

$$D\varphi(\zeta)\,\zeta^{-k} = O(1) \quad \text{as} \quad \zeta \longrightarrow 0,$$

and from (7.5) that

$$\lim_{\substack{\zeta\to 0\\ \zeta\neq 0}} \frac{\varphi_z(\zeta)}{\zeta^k}$$

exists. This limit is zero because the asymptotic expansion

$$\varphi_z(\zeta) = A\,\zeta^k + o(|\zeta|^k)$$

gives, by the mean value theorem,

$$\begin{aligned}\varphi(z) &= \int_0^1 D\varphi(\tau z)\,d\tau\cdot z\\ &= \int_0^1 (\varphi_x(\tau z)x + \varphi_y(\tau z)y)\,d\tau\\ &= 2\,\mathrm{Re}\int_0^1 \varphi_z(\tau z)\,z\,d\tau\\ &= 2\,\mathrm{Re}\int_0^1 \{A(\tau z)^k z + o(|\tau z|^k)z\}\,d\tau\\ &= \mathrm{Re}\left[\frac{2A}{k+1}z^{k+1}\right] + o(|z|^{k+1}),\end{aligned}$$

from which $A = 0$ since $k < n$.

The induction is complete, i.e. (7.4) is true for $k = n$. By arguing as above one can now deduce the existence of the limit

$$\lim_{\substack{\zeta\to 0\\ \zeta\neq 0}} \frac{\varphi_z(\zeta)}{\zeta^n}. \ \square$$

7.2. Generalization to differential inequalities with Hölder continuous coefficients

Theorem 7.2.1. *Suppose that* $\varphi(z) \in C^1(\Omega)$ *satisfies the differential inequality*

$$|D_\alpha(a(z)\,D_\alpha\varphi)| \le C(|D\varphi| + |\varphi|), \tag{7.8}$$

where $a \in C^\mu(\Omega)$, $a > 0$, *for some* μ, $0 < \mu < 1$. *If* $\varphi(z) = o(|z|^n)$ *as* $|z| \to 0$ *for some* $n \in \mathbb{N}_0$, *then*

$$\lim_{\substack{|z|\to 0\\ z\neq 0}} \frac{\varphi_z(z)}{z^n} \tag{7.9}$$

exists.

An approximation lemma will first be established, since the Carleman–Hartman–Wintner theorem extends to the case $a \in C^1$:

Lemma 7.2.2. *Let* $a \in C^\mu(\Omega)$ *for some* μ, $0 < \mu < 1$, *which satisfies*

$$0 < \lambda \le a(z) \le \Lambda < +\infty, \tag{7.10}$$

$$[a]_\mu^\Omega \le H. \tag{7.11}$$

Then there exists a disc $B = B_{R_0}(0)$, $0 < R_0 = R_0(\mu, \lambda, \Lambda, H) < \mathrm{dist}(0, \partial\Omega)$, *such that for each* ζ, $0 < |\zeta| < R_0$, *there is a function* $\tilde{a} \in C^{\mu}(B) \cap C^1(B\backslash\{0\})$ *such that*

$$\tilde{a}(0) = a(0), \qquad \tilde{a}(\zeta) = a(\zeta), \tag{7.12}$$

$$\frac{\lambda}{2} \le \tilde{a}(z) \le 2\Lambda \qquad (z \in B), \tag{7.13}$$

$$[\tilde{a}]_{\mu}^{B} \le C, \tag{7.14}$$

$$|D\tilde{a}(z)| \le C\,|z|^{\mu-1} \qquad (z \neq 0), \tag{7.15}$$

where $C = C(\mu, \lambda, \Lambda, H, \mathrm{dist}(0, \partial\Omega))$.

<u>Proof:</u> For $\epsilon > 0$ set

$$\varphi_{\epsilon}(z) = \int_{\Omega} \phi_{\epsilon}(\zeta - z)\, a(\zeta)\, d\xi\, d\eta.$$

Here

$$\phi_{\epsilon}(z) = \frac{1}{\epsilon^2}\phi\!\left(\frac{z}{\epsilon}\right),$$

where $\phi \in C^{\infty}$ such that $\phi \ge 0$, $\int \phi = 1$, $\phi(z) = 0$ for $|z| \ge 1$. Then let

$$a_{\epsilon}(z) = \varphi_{\epsilon}(z) + \frac{(a(\zeta) - \varphi_{\epsilon}(\zeta)) \; - \; (a(0) - \varphi_{\epsilon}(0))}{|\zeta|^{\mu}}\,|z|^{\mu} + (a(0) - \varphi_{\epsilon}(0)).$$

If R_0, ϵ_0 are sufficiently small, depending only on μ, λ, Λ, H, $\mathrm{dist}(0, \partial\Omega)$), then

$$\tilde{a}(z) = a_{\epsilon_0}(z)$$

satisfies all the stated properties. □

<u>Proof of Theorem 7.2.1:</u> Assume that $\varphi \in C^2$ for the sake of exposition and rewrite (7.8) in the (weak) form

$$-D_{\alpha}(\tilde{a}\, D_{\alpha}\varphi) = D_{\alpha}((a - \tilde{a})\, D_{\alpha}\varphi) + W.$$

Compute

$$\begin{aligned} 4(\tilde{a}\,\varphi_z)_{\bar{z}} &= 2\,((\tilde{a}\,\varphi_z)_x + i\,(\tilde{a}\varphi_z)_y) \\ &= (\tilde{a}\,(\varphi_x - i\,\varphi_y))_x + i\,(\tilde{a}\,(\varphi_x - i\,\varphi_y))_y \\ &= (\tilde{a}\,\varphi_x)_x + (\tilde{a}\,\varphi_y)_y + i\,(\varphi_x\,\tilde{a}_y - \varphi_y\,\tilde{a}_x). \end{aligned}$$

Therefore

$$-4(\tilde{a}\,\varphi_z)_{\bar{z}} = D_\alpha((a-\tilde{a})\,D_\alpha\varphi) + i\,(\tilde{a}_x\,\varphi_y - \tilde{a}_y\,\varphi_x) + W$$
$$= \tilde{W}\,.$$

Multiplying by

$$\eta(z) = z^{-k}\,(z-\zeta)^{-1}$$

$(k = 1, 2, \ldots, n)$, and integrating over $\Omega_\epsilon = B_R \backslash (B_\epsilon \cup B_\epsilon(\zeta))$ yields

$$2i\int_{\partial\Omega_\epsilon} \tilde{a}\,\varphi_z\,\eta\,dz = \int_{\Omega_\epsilon} \tilde{W}\,\eta\,dx\,dy\,.$$

If

(7.16) $$\varphi_z = o(|z|^{k-1}) \quad \text{as} \quad |z| \longrightarrow 0\,,$$

then one can let $\epsilon \longrightarrow 0$ to obtain

(7.17) $$4\pi\,a(\zeta)\,\varphi_z(\zeta)^{-k} = -2i\int_{\partial B_R} \tilde{a}\,\varphi_z\,\eta\,dz + \int_{\partial B_R} (a-\tilde{a})\,D_\nu\varphi\,\eta\,ds$$
$$+ \int_{B_R} \{(\tilde{a}-a)\,D_\alpha\varphi D_\alpha\eta + i\,(\tilde{a}_x\,\varphi_y - \tilde{a}_y\,\varphi_x)\,\eta + W\,\eta\}\,dx\,dy\,.$$

Again, one shows by induction over k, $1 \le k \le n$, that (7.16) holds: Suppose that (7.16) holds for a k, $1 \le k < n$. By Lemma 7.2.2, all integrals in (7.17) are absolutely convergent. Hence there exists a C such that

(7.18) $$|D\varphi(\zeta)|\,|\zeta|^{-k} \le C\Big\{\int_{\partial B_R} |D\varphi|\,|z|^{-k}\,|z-\zeta|^{-1}\,ds + \int_{B_R} |\varphi|\,|z|^{-k}\,|z-\zeta|^{-1}\,dx\,dy$$
$$+ \int_{B_R} |D\varphi|\,(|z|^{-k-1+\mu}\,|z-\zeta|^{-1} + |z|^{-k}\,|z-\zeta|^{-2+\mu}\,dx\,dy\Big\}\,.$$

This inequality is multiplied by $|\zeta-z_0|^{-2+\mu}$ and integrated over B_R. By virtue of

$$|(z-\zeta)(\zeta-z_0)|^{-\gamma} \le C(\gamma)\,|z-z_0|^{-\gamma}(|z-\zeta|^{-\gamma} + |\zeta-z_0|^{-\gamma})$$

for $\gamma > 0$, one estimates

$$\int_{B_R} |D\varphi(\zeta)|\,|\zeta|^{-k}\,|\zeta-z_0|^{-2+\mu}d\xi\, d\eta$$

$$\le C\Big\{\int_{\partial B_R} |D\varphi|\,|z|^{-k}\int_{B_R} |z-\zeta|^{-1}\,|\zeta-z_0|^{-2+\mu}d\xi\, d\eta\, ds$$

$$+\int_{B_R} |\varphi|\,|z|^{-k}\int_{B_R} |z-\zeta|^{-1}\,|\zeta-z_0|^{-2+\mu}d\xi\, d\eta\, dx\, dy$$

$$+\int_{B_R} |D\varphi|\,|z|^{-k-1+\mu}\int_{B_R} |z-\zeta|^{-1}\,|\zeta-z_0|^{-2+\mu}d\xi\, d\eta\, dx\, dy$$

$$+\int_{B_R} |D\varphi|\,|z|^{-k}\int_{B_R} |z-\zeta|^{-2+\mu}\,|\zeta-z_0|^{-2+\mu}d\xi\, d\eta\, dx\, dy\Big\}$$

$$\le C R^{\mu}\Big\{\int_{\partial B_R} |D\varphi|\,|z|^{-k}\,|z-z_0|^{-1}\, ds + \int_{B_R} |\varphi|\,|z|^{-k}\,|z-z_0|^{-1}\, dx\, dy$$

$$+\int_{B_R} |D\varphi|\,|z|^{-k-1+\mu}\,|z-z_0|^{-1}\, dx\, dy + \int_{B_R} |D\varphi|\,|z|^{-k}\,|z-z_0|^{-2+\mu}dx\, dy\Big\}.$$

After relabelling one gets

$$(7.19)\quad (1-CR^{\mu})\int_{B_R} |D\varphi|\,|z|^{-k}|z-\zeta|^{-2+\mu}dx\, dy \le C R^{\mu}\Big\{\int_{\partial B_R} |D\varphi|\,|z|^{-k}\,|z-\zeta|^{-1}\, ds$$

$$+\int_{B_R} |\varphi|\,|z|^{-k}\,|z-\zeta|^{-1}\, dx\, dy + \int_{B_R} |D\varphi|\,|z|^{-k-1+\mu}\,|z-\zeta|^{-1}\, dx\, dy\Big\}.$$

Remark 7.2.3. Note that at this point it is not clear at all whether the proof will work out, because in (7.19) we have estimated one integral of the r. h. s. of (7.18) in terms of the other. It is of course not enough to proceed estimating the latter integral in terms of the former. Surprisingly, we can break the circle as follows:

Proof of Theorem 7.2.1 continued: Now multiply (7.18) by $|\zeta|^{-1+\mu}|\zeta-z_0|^{-1}$ and integrate over B_R to obtain

$$\int_{B_R} |D\varphi(\zeta)|\,|\zeta|^{-k-1+\mu}\,|\zeta-z_0|^{-1}\, d\xi\, d\eta$$

$$\le C\Big\{\int_{\partial B_R} |D\varphi|\,|z|^{-k}\int_{B_R} |z-\zeta|^{-1}\,|\zeta|^{-1+\mu}\,|\zeta-z_0|^{-1}\, d\xi\, d\eta\, ds$$

$$+\int_{B_R} |\varphi|\,|z|^{-k}\int_{B_R} |z-\zeta|^{-1}\,|\zeta|^{-1+\mu}\,|\zeta-z_0|^{-1}\, d\xi\, d\eta\, dx\, dy$$

$$+\int_{B_R} |D\varphi|\,|z|^{-k-1+\mu}\int_{B_R} |z-\zeta|^{-1}\,|\zeta|^{-1+\mu}\,|\zeta-z_0|^{-1}\, d\xi\, d\eta\, dx\, dy$$

$$+\int_{B_R} |D\varphi|\,|z|^{-k}\int_{B_R} |z-\zeta|^{-2+\mu}\,|\zeta|^{-1+\mu}\,|\zeta-z_0|^{-1}\, d\xi\, d\eta\, dx\, dy\Big\}.$$

Now estimate

$$\int_{B_R} |z-\zeta|^{-1} |\zeta|^{-1+\mu} |\zeta - z_0|^{-1} d\xi\, d\eta$$

$$\leq C|z-z_0|^{-1} \int_{B_R} |\zeta|^{-1+\mu} (|z-\zeta|^{-1} + |\zeta - z_0|^{-1}) d\xi\, d\eta$$

$$\leq C|z-z_0|^{-1} R^{\mu},$$

and

$$\int_{B_R} |z-\zeta|^{-2+\mu} |\zeta|^{-1+\mu} |\zeta - z_0|^{-1} d\xi\, d\eta$$

$$\leq C|z-z_0|^{-1} \int_{B_R} |z-\zeta|^{-1+\mu} |\zeta|^{-1+\mu} (|z-\zeta|^{-1} + |\zeta - z_0|^{-1}) d\xi\, d\eta$$

$$\leq C|z-z_0|^{-1} |z|^{-1+\mu} \int_{B_R} (|z-\zeta|^{-1+\mu} + |\zeta|^{-1+\mu})(|z-\zeta|^{-1} + |\zeta - z_0|^{-1}) d\xi\, d\eta$$

$$\leq C|z-z_0|^{-1} |z|^{-1+\mu} R^{\mu}.$$

Therefore

$$(1 - CR^{\mu}) \int_{B_R} |D\varphi|\, |z|^{-k-1+\mu} |z-\zeta|^{-1} dx\, dy$$

$$\leq CR^{\mu} \left\{ \int_{\partial B_R} |D\varphi|\, |z|^{-k} |z-\zeta|^{-1} ds + \int_{B_R} |\varphi|\, |z|^{-k} |z-\zeta|^{-1} dx\, dy \right\}. \tag{7.20}$$

If $R^{\mu} \leq 1/2C$, then (7.19) and (7.20) can be combined to give

$$\int_{B_R} |D\varphi| (|z|^{-k-1+\mu} |z-\zeta|^{-1} + |z|^{-k} |z-\zeta|^{-2+\mu}) dx\, dy$$

$$\leq CR^{\mu} \left\{ \int_{\partial B_R} |D\varphi|\, |z|^{-k} |z-\zeta|^{-1} ds + \int_{B_R} |\varphi|\, |z|^{-k} |z-\zeta|^{-1} dx\, dy \right\}. \tag{7.21}$$

The r. h. s. is $O(1)$ as $\zeta \to 0$. Hence, by (7.18),

$$D\varphi(\zeta)\, \zeta^{-k} = O(1),$$

and by (7.17) it follows that

$$\lim_{\substack{\zeta \to 0 \\ \zeta \neq 0}} \frac{\varphi_z(\zeta)}{\zeta^{-k}} \quad \text{exists}. \tag{7.22}$$

Finally, if $k < n$, then this limit is zero, which completes the induction, and the same argument yields the existence of the limit (7.22) for $k = n$ as required. □

<u>Theorem 7.2.4.</u> *Suppose that* $\varphi(z) \in C^1(\Omega)$ *satisfies the elliptic inequality*

$$|D_\alpha(a^{\alpha\beta}(z)\,D_\beta\varphi)| \le C(|D\varphi| + |\varphi|) \tag{7.23}$$

in the weak sense. The coefficient matrix $[a^{\alpha\beta}(z)]$ *is assumed to be real, symmetric, positive definite and* $a^{\alpha\beta} \in C^\mu(\Omega)$ *for some* μ, $0 < \mu < 1$. *W.l.o.g., we make the normalization* $a^{11}(0) = a^{22}(0)$, $a^{12}(0) = a^{21}(0) = 0$. *Then the statement of Theorem* 7.2.1 *remains true.*

<u>Proof:</u> It is convenient to rewrite the coefficients $a^{\alpha\beta}$ in the form

$$a^{\alpha\beta} = a\,A^{\alpha\beta}, \qquad a^2 = \Delta = \det[a^{\alpha\beta}].$$

Consider the differential form

$$A^{22}\,dx^2 - 2\,A^{12}\,dx\,dy + A^{11}\,dy^2 = A_{\alpha\beta}dx^\alpha dx^\beta.$$

According to Remark 6.4.1, one can find a (local) diffeomorphism $w = (u,v) = (u^1,u^2)$ of class $C^{1,\mu}$ satisfying the conformality relations (6.63, 64),

$$A^{\alpha\beta}\frac{\partial u}{\partial x^\alpha}\frac{\partial u}{\partial x^\beta} = A^{\alpha\beta}\frac{\partial v}{\partial x^\alpha}\frac{\partial v}{\partial x^\beta} = Jw,$$

$$A^{\alpha\beta}\frac{\partial u}{\partial x^\alpha}\frac{\partial v}{\partial x^\beta} = 0.$$

Hence if φ and η are C^1–functions, then

$$\begin{aligned} A^{\alpha\beta}D_\alpha\varphi D_\beta\eta &= A^{\alpha\beta}\frac{\partial u^i}{\partial x^\alpha}\frac{\partial u^j}{\partial x^\beta}D_i\varphi D_j\eta \\ &= Jw(\varphi_u\eta_u + \varphi_v\eta_v), \end{aligned}$$

and therefore

$$\begin{aligned} \int_{w(\Omega)} a(\varphi_u\eta_u + \varphi_v\eta_v)\,du\,dv &= \int_\Omega a\,A^{\alpha\beta}D_\alpha\varphi D_\beta\eta\,dx\,dy \\ &= \int_\Omega W\,dx\,dy \\ &= \int_{w(\Omega)} \tilde{W}\,du\,dv. \end{aligned}$$

Theorem 7.2.1 is therefore applicable, yielding the existence of the limit

$$\lim_{\substack{|w|\to 0\\ w\neq 0}} \frac{\varphi_w(w)}{w^n}.$$

In order to translate this back to the $z = x+iy$ variable, one uses the normalization $A^{11}(0) = A^{22}(0)$, $A^{12}(0) = 0$, i.e.,

$$u_x(0) = v_y(0), \quad u_y(0) = v_x(0) = 0. \ \square$$

7.3. The unique continuation principle

The following theorem is a well-known unique continuation principle, which is true for even more general operators than stated.

Theorem 7.3.1. *Suppose that* Ω *is a domain. Let* $\varphi(z) \in C^1(\Omega)$ *satisfy the elliptic inequality*

$$|D_\alpha(a^{\alpha\beta}(z)\, D_\beta\varphi)| \le C(|D\varphi| + |\varphi|), \tag{7.24}$$

such that the coefficients $a^{\alpha\beta}$ *satisfy the assumptions of Theorem* 7.2.4. *If* $\varphi = o(|z|^n)$ as $|z| \longrightarrow 0$ *for all* $n \in \mathbb{N}$, *then* $\varphi(z) \equiv 0$.

Proof: For the differential inequality (7.8),

$$|D_\alpha(a(z)\, D_\alpha\varphi)| \le C(|D\varphi| + |\varphi|),$$

we have derived the inequality (7.21),

$$\int_{B_R} |D\varphi|\, |\, |z|^{-k-1+\mu}|z-\zeta|^{-1} + |z|^{-k}|z-\zeta|^{-2+\mu})\, dx\, dy$$

$$\le CR^\mu\Big\{\int_{\partial B_R} |D\varphi|\, |z|^{-k}|z-\zeta|^{-1} ds + \int_{B_R} |\varphi|\, |z|^{-k}|z-\zeta|^{-1} dx\, dy\Big\}$$

if $CR^\mu \le 1/2$. By letting $\zeta \longrightarrow 0$,

$$\int_{B_R} |D\varphi|\, |z|^{-k-2+\mu} dx\, dy \le CR^\mu\Big\{\int_{\partial B_R} |D\varphi|\, |z|^{-k-1} ds + \int_{B_R} |\varphi|\, |z|^{-k-1} dx\, dy\Big\}.$$

Now

$$\varphi(z) = \int_0^1 \frac{d}{d\tau}(\varphi((1-\tau)0 + \tau z))\, d\tau$$

$$= \int_0^1 D\varphi(\tau z)\, d\tau \cdot z,$$

hence

$$\int_{B_R} |\varphi|\,|z|^{-k-1}\,dx\,dy \le \int_0^1\int_{B_R} |D\varphi(\tau z)|\,|z|^{-k}\,dx\,dy\,d\tau$$

$$= \int_0^1 \tau^{k-2}\int_{B_R} |D\varphi(z)|\,|z|^{-k}\,dx\,dy\,d\tau$$

$$\le \int_{B_R} |D\varphi(z)|\,|z|^{-k}\,dx\,dy\,,$$

if $k \ge 2$. Therefore, by swallowing,

$$\int_{B_R} |D\varphi|\,|z|^{-k-2+\mu}\,dx\,dy \le C R^{\mu}\int_{\partial B_R} |D\varphi|\,|z|^{-k-1}\,ds\,.$$

Suppose that there exists a point $z_0 \in B_R(0)$ for which $D\varphi(z_0) \ne 0$. Then $|D\varphi(z)| \ge \frac{1}{2}|D\varphi(z_0)|$ for $|z-z_0| \le \epsilon$, and therefore

$$|D\varphi(z_0)|\,|z_0|^{-k-2+\mu} \le C(\epsilon, z_0)\,R^{\mu}\int_{\partial B_R} |D\varphi|\,|z|^{-k-1}\,ds$$

$$\le C R^{-k}.$$

Hence

$$|D\varphi(z_0)| \le C\left[\frac{|z_0|}{R}\right]^k .$$

Since $|z_0| < R$, it follows that $D\varphi(z_0) = 0$ by letting $k \longrightarrow \infty$, a contradiction. It follows that $\varphi(z) \equiv 0$.

For the general case (7.24) one utilizes a conformal mapping as in the proof of Theorem 7.2.4. □

7.4. Local behavior of solutions of differential inequalities

In this section $\varphi(z) \in C^1(\Omega)$ is a solution of the elliptic inequality

(7.25) $$|D_\alpha(a^{\alpha\beta} D_\beta\varphi)| \le C(|D\varphi| + |\varphi|)\,.$$

Theorem 7.4.1. *Let* $\varphi(z) = o(|z|^n)$ *as* $|z| \longrightarrow 0$ *for some* $n \in \mathbb{N}$. *Then either* $\varphi(z) \equiv 0$ *or there exists a* $m \ge n$, $m \in \mathbb{N}$ *such that*

(7.26) $$\lim_{\substack{|z|\to 0\\ z\ne 0}} \frac{\varphi_z(z)}{z^m} \ne 0\,.$$

Proof: Let

$$m = \sup\{n \in \mathbb{N} \mid \varphi(z) = o(|z|^n) \text{ as } |z| \to \infty\}$$

if $\varphi(z) \not\equiv 0$. Then $m \geq n$ and $m < +\infty$ by the unique continuation principle, Theorem 7.3.1. The existence of

$$\lim_{\substack{|z|\to 0 \\ z\neq 0}} \frac{\varphi_z(z)}{z^m} = A'$$

follows from Theorem 7.2.4. Hence $\varphi_z(z) = A' z^m + o(|z|^m)$. By the mean value theorem,

$$\begin{aligned}
\varphi(z) &= \int_0^1 \{\varphi_x(\tau z)x + \varphi_y(\tau z)y\}\,d\tau \\
&= 2\,\mathrm{Re}\int_0^1 \varphi_z(\tau z)\,z\,d\tau \\
&= 2\,\mathrm{Re}\int_0^1 \{A'(\tau z)^m z + o(|\tau z|^m)z\}\,d\tau \\
&= \mathrm{Re}\left[\frac{2A'}{m+1}z^{m+1}\right] + o(|z|^{m+1}).
\end{aligned}$$

This implies that $A' \neq 0$. □

Corollary 7.4.2. *If* $\varphi(z) \not\equiv 0$, *then* $\varphi(z)$ *has an asymptotic expansion of the form*

$$\varphi(z) = \mathrm{Re}(A z^{m+1}) + o(|z|^{m+1}) \tag{7.27}$$

as $|z| \to 0$, *where* $m \geq n$ *and*

$$A = \frac{2}{m+1} \lim_{\substack{|z|\to 0 \\ z\neq 0}} \frac{\varphi_z(z)}{z^m} \neq 0. \tag{7.28}$$

Proposition 7.4.3. *Let* $\{\varphi^{(k)}(z)\}_{k=1}^{\infty}$ *be a sequence of functions of class* $C^1(\Omega)$ *satisfying the differential inequality* (7.25), *where* C *is independent of* k. *Assume that*

$$\varphi^{(k)}(z) \to \varphi(z), \qquad D\varphi^{(k)}(z) \to \varphi(z) \tag{7.29}$$

uniformly in Ω $(k \to \infty)$. *Let* $\varphi(z) = o(|z|)$ *as* $|z| \to 0$ *and assume that* $D\varphi^{(k)}(z) \neq 0$ *in* Ω *for all* $k \in \mathbb{N}$. *Then* $\varphi(z) \equiv 0$.

<u>**Proof:**</u> If $\varphi \not\equiv 0$, then, by Theorem 7.4.1,

$$\lim_{\substack{|z|\to 0 \\ z\neq 0}} \frac{\varphi_z(z)}{z^m} = A \neq 0,$$

i.e.,

$$\varphi_z(z) = A z^m + o(|z|^m).$$

Put $\psi(z) = A z^m$. Then

$$|\varphi_z(z) - \psi(z)| = o(|z|^m) \leq \tfrac{1}{2}|\psi(z)|$$

for $|z| \leq R_0$, R_0 small. Let $|z| = R_0$, then

$$\begin{aligned} |\varphi_z^{(k)}(z) - \psi(z)| &\leq |\varphi_z^{(k)}(z) - \varphi_z(z)| + \tfrac{1}{2}|\psi(z)| \\ &\leq |\psi(z)| \end{aligned}$$

if $k \geq k_0$. By the homotopy invariance of the winding number,

$$\frac{1}{2\pi} \oint_{|z|=R_0} d(\arg \varphi_z^{(k)}(z)) = \frac{1}{2\pi} \int_{|z|=R_0} d(\arg \psi(z)) = m > 0.$$

$\varphi_z^{(k)}$ must therefore have a zero in B_{R_0}, a contradiction. □

Chapter 8. UNIVALENT SOLUTIONS OF HEINZ–LEWY TYPE SYSTEMS

8.1. Elliptic systems of Heinz–Lewy type

Consider a solution $w = (u,v) = w(x,y) = w(z) \in C^1(\Omega)$ of the Heinz–Lewy type system

$$
\begin{aligned}
Lu &= h_1(w)\,|Du|^2 + h_2(w)\,Du\cdot Dv + h_3(w)\,|Dv|^2 + h_4(w)\,Du\wedge Dv,\\
Lv &= \tilde h_1(w)\,|Du|^2 + \tilde h_2(w)\,Du\cdot Dv + \tilde h_3(w)\,|Dv|^2 + \tilde h_4(w)\,Du\wedge Dv,
\end{aligned}
\tag{8.1}
$$

where

$$
\begin{aligned}
L &= -\frac{1}{a(z,w)}D_\alpha(a(z,w)D_\alpha)\\
&= -\frac{1}{a(z,w)}\left[\frac{\partial}{\partial x}\left[a(z,w)\frac{\partial}{\partial x}\right]\right] + \frac{\partial}{\partial y}\left[a(z,w)\frac{\partial}{\partial y}\right]\Big].
\end{aligned}
\tag{8.2}
$$

With regards to the coefficients $a, h_1, \ldots, \tilde h_4$, we make the following

Assumption (A8.1). (i) $a \in C^\mu(\Omega \times \mathbb{R}^2)$ for some exponent μ, $0 < \mu < 1$, such that

$$0 < \lambda \le a(z,w) \le \Lambda < +\infty \tag{8.3}$$

for all $(z,w) \in \Omega \times \mathbb{R}^2$ and

$$[a]_\mu^{\Omega\times\mathbb{R}^2} \le H. \tag{8.4}$$

(ii) The coefficients $h_1, \ldots, \tilde h_4$ are bounded, Borel measurable functions on $\mathbb{R}^2$ such that

$$|h_1(w)|, \ldots, |\tilde h_4(w)| \le M_0 \tag{8.5}$$

for all $w \in \mathbb{R}^2$, and the functions

$$
\begin{aligned}
\omega_1(w) &= \tilde h_1(w),\\
\omega_2(w) &= h_1(w) - \tilde h_2(w),\\
\omega_3(w) &= h_2(w) - \tilde h_3(w),\\
\omega_4(w) &= h_3(w)
\end{aligned}
\tag{8.6}
$$

$(w \in \mathbb{R}^2)$ are Lipschitz continuous with Lipschitz constant $\le M_1$:

$$[\omega_1]_{0,1}^{\mathbb{R}^2}, \ldots, [\omega_4]_{0,1}^{\mathbb{R}^2} \le M_1. \tag{8.7}$$

<u>Lemma 8.1.1.</u> *Let* T *be an orthogonal matrix,*

(8.8)
$$T=\begin{bmatrix} \xi & \eta \\ -\eta & \xi \end{bmatrix}, \qquad \xi^2+\eta^2=1.$$

Then $\hat{w}(z)=Tw(z)$ *solves a system of the form* (8.1) *with coefficients* $\hat{h}_1,\dots,\hat{\tilde{h}}_4$, *which satisfy the Assumption* (A8.1) *with constants* $\hat{M}_0=\hat{M}_0(M_0)$, $\hat{M}_1=\hat{M}_1(M_1)$.

<u>Proof:</u>

$$\hat{u}=\xi u+\eta v,$$
$$\hat{v}=-\eta u+\xi v,$$

$$u=\xi\hat{u}-\eta\hat{v},$$
$$v=\eta\hat{u}+\xi\hat{v},$$

hence

$$\begin{aligned}
L\hat{u} &= \xi Lu+\eta Lv \\
&= (\xi h_1+\eta\tilde{h}_1)\,|Du|^2+(\xi h_2+\eta\tilde{h}_2)\,Du\cdot Dv \\
&\quad +(\xi h_3+\eta\tilde{h}_3)\,|Dv|^2+(\xi h_4+\eta\tilde{h}_4)\,Du\wedge Dv \\
&= (\xi h_1+\eta\tilde{h}_1)(\xi^2|D\hat{u}|^2-2\xi\eta D\hat{u}D\hat{v}+\eta^2|D\hat{v}|^2) \\
&\quad +(\xi h_2+\eta\tilde{h}_2)(\xi\eta|D\hat{u}|^2+(\xi^2-\eta^2)D\hat{u}D\hat{v}-\xi\eta|D\hat{v}|^2) \\
&\quad +(\xi h_3+\eta\tilde{h}_3)(\eta^2|D\hat{u}|^2+2\xi\eta D\hat{u}D\hat{v}+\xi^2|D\hat{v}|^2)+(\xi h_4+\eta\tilde{h}_4)\,D\hat{u}\wedge D\hat{v} \\
&= (\xi^3h_1+\xi^2\eta(\tilde{h}_1+h_2)+\xi\eta^2(\tilde{h}_2+h_3)+\eta^3\tilde{h}_3)\,|D\hat{u}|^2 \\
&\quad +(\xi^3h_2+\xi^2\eta(-2h_1+\tilde{h}_2+2h_3)+\xi\eta^2(-2\tilde{h}_1-h_2+2\tilde{h}_3)-\eta^3\tilde{h}_2)\,D\hat{u}\cdot D\hat{v} \\
&\quad +(\xi^3h_3+\xi^2\eta(\tilde{h}_3-h_2)+\xi\eta^2(h_1-\tilde{h}_2)+\eta^3\tilde{h}_1)\,|D\hat{v}|^2+(\xi h_4+\eta\tilde{h}_4)\,D\hat{u}\wedge D\hat{v} \\
&= \hat{h}_1\,|D\hat{u}|^2+\hat{h}_2\,D\hat{u}\cdot D\hat{v}+\hat{h}_3\,|D\hat{v}|^2+\hat{h}_4\,D\hat{u}\wedge D\hat{v}.
\end{aligned}$$

Similarly,

$$\begin{aligned}
L\hat{v} &= (\xi^3\tilde{h}_1+\xi^2\eta(-h_1+\tilde{h}_2)+\xi\eta^2(-h_2+\tilde{h}_3)-\eta^3h_3)\,|D\hat{u}|^2 \\
&\quad +(\xi^3\tilde{h}_2+\xi^2\eta(-2\tilde{h}_1-h_2+2\tilde{h}_3)+\xi\eta^2(2h_1-\tilde{h}_2-2h_3)+\eta^3h_2)\,D\hat{u}\cdot D\hat{v} \\
&\quad +(\xi^3\tilde{h}_3+\xi^2\eta(-\tilde{h}_2-h_3)+\xi\eta^2(\tilde{h}_1+h_2)-\eta^3h_1)\,|D\hat{v}|^2+(\xi\tilde{h}_4-\eta h_4)\,D\hat{u}\wedge D\hat{v} \\
&= \hat{\tilde{h}}_1\,|D\hat{u}|^2+\hat{\tilde{h}}_2\,D\hat{u}\cdot D\hat{v}+\hat{\tilde{h}}_3\,|D\hat{v}|^2+\hat{\tilde{h}}_4\,D\hat{u}\wedge D\hat{v}.
\end{aligned}$$

The Lipschitz continuity of $\hat{\tilde{h}}_1$, $\hat{h}_1-\hat{\tilde{h}}_2$, $\hat{h}_2-\hat{\tilde{h}}_3$, $\hat{h}_3$ follows by inspection. □

Proposition 8.1.2. *There exists a disc* $B_\delta = B_\delta(0)$ *in the* w*-plane,* $\delta = \delta(M_0)$*, such that the following statement is true: For each* $\zeta = (\xi, \eta) \in \mathbb{R}^2$*,* $|\zeta| = 1$*, there is a function* $\phi \in C^2(B_\delta)$ *with* $\phi(0) = 0$*,* $\phi_u(0) = \xi$*,* $\phi_v(0) = \eta$*,* $\|\phi\|_{C^{1,1}} \le C(M_0)$*,*

$$\phi(u,v) = u\xi + v\eta - g(v\xi - u\eta), \tag{8.9}$$

and such that $\varphi(z) = \phi(w(z))$ *satisfies a differential inequality of the form*

$$|L\varphi| \le C(M_0, M_1, K)(|D\varphi| + |\varphi|) \tag{8.10}$$

in B_R *for any* $w \in C^1(B_R, B_\delta)$ *solving the system* (8.1) *with* $w(0) = 0$*,* $|Dw| \le K$*.*

Proof: Assume that $w \in C^2$ for convenience only and let $g = g(t) \in C^2$ be real valued. Consider

$$\varphi(z) = \phi(w(z)) = u(z) - g(v(z))$$

and calculate

$$\begin{aligned} L\varphi &= Lu + \tfrac{1}{a} D_\alpha(a g' D_\alpha v) \\ &= Lu - g' Lv + g'' |Dv|^2 \\ &= (h_1 - g' \tilde{h}_1)|Du|^2 + (h_2 - g' \tilde{h}_2) Du \cdot Dv \\ &\quad + (h_3 - g' \tilde{h}_3 + g'')|Dv|^2 + (h_4 - g' \tilde{h}_4) Du \wedge Dv. \end{aligned}$$

Now use the fact that

$$Du = D\varphi + g' Dv$$

to compute

$$\begin{aligned} L\varphi = {} & (h_1 - g'\tilde{h}_1)|D\varphi|^2 + (2(h_1 - g'\tilde{h}_1)g' + (h_2 - g'\tilde{h}_2)) D\varphi \cdot Dv \\ & + ((h_1 - g'\tilde{h}_1)(g')^2 + (h_2 - g'\tilde{h}_2)g' + (h_3 - g'\tilde{h}_3 + g''))|Dv|^2 \\ & + (h_4 - g'\tilde{h}_4) D\varphi \wedge Dv \\ = {} & (h_1 - \tilde{h}_1 g')|D\varphi|^2 + (h_2 + (2h_1 - \tilde{h}_2)g' - 2\tilde{h}_1 (g')^2) D\varphi \cdot Dv \\ & + (h_3 + (h_2 - \tilde{h}_3)g' + (h_1 - \tilde{h}_2)(g')^2 - \tilde{h}_1 (g')^3 + g'')|Dv|^2 \\ & + (h_4 - \tilde{h}_4 g') D\varphi \wedge Dv \\ = {} & A(z)\varphi_x + B(z)\varphi_y + \Gamma(u, v, g', g'')|Dv|^2, \end{aligned}$$

where $A(z)$, $B(z)$ are certain expressions in $h_1, \ldots, \tilde{h}_4$, g', $D\varphi$, Dv, and

$$\Gamma(u,v,g',g'') = \omega_4(u,v) + \omega_3(u,v)\,g' + \omega_2(u,v)\,(g')^2 - \omega_1(u,v)\,(g')^3 + g''.$$

Let

$$f(t,g_0,g_1) = -\omega_4(g_0,t) - \omega_3(g_0,t)\,g_1 - \omega_2(g_0,t)\,g_1^2 + \omega_1(g_0,t)\,g_1^3.$$

Then f is locally Lipschitz continuous in $\mathbb{R}^3$ and

$$\sup_{\mathbb{R}^2 \times [-1,1]} |f| \le C(M_0).$$

One can therefore solve the differential equation

$$g'' = f(t,g,g') \quad \text{for} \quad |t| \le \delta = \delta(M_0)$$

subject to the initial condition

$$g(0) = g'(0) = 0$$

such that

$$|g'(t)| \le 1.$$

It follows that

$$|g''(t)| \le C(M_0).$$

Therefore

$$\Gamma(u,v,g',g'') = (\omega_4(u,v) - \omega_4(g(v),v)) + (\omega_3(u,v) - \omega_3(g(v),v))\,g'(v)$$
$$+ (\omega_2(u,v) - \omega_2(g(v),v))\,(g'(v))^2 - (\omega_1(u,v) - \omega_1(g(v),v))\,(g'(v))^3,$$

which implies

$$|\Gamma(u,v,g',g'')| \le C(M_1)\,|u - g(v)|$$
$$\le C\,|\varphi|,$$

and therefore

$$|L\varphi| \le C(|D\varphi| + |\varphi|).$$

The conditions $\phi_u(0) = \xi$, $\phi_v(0) = \eta$ can be achieved by taking $\hat{w} = Tw$, T an orthogonal matrix, instead of w:

$$\Phi(w) = \phi(Tw) = (\xi u + \eta v) + g(-\eta u + \xi v). \quad \square$$

8.2. Non–vanishing of the Jacobian

The following topological lemma is due to Berg [BG]:

Lemma 8.2.1. *Let* $\Phi(z) \in C^0(B_R(0), \mathbb{C})$ *be a homeomorphism from* $B_R(0)$ *onto* $\Omega \subset \mathbb{C}$. *Assume that* $\varphi(z) = \mathrm{Re}\,\Phi(z)$ *has an asymptotic expansion of the form*

(8.11)
$$\varphi(z) = \mathrm{Re}(A\,z^{m+1}) + o(|z|^{m+1})$$

as $|z| \longrightarrow 0$ *with* $A \neq 0$, $m \in N_0$. *Then* $m = 0$.

Proof: We can assume that $\Phi(0) = 0$ and that $A \in \mathbb{R}$, $A > 1$, because if $A = |A|\,e^{i\vartheta}$, then let

$$z = a\,\zeta, \qquad a = \left[\frac{\alpha}{|A|}e^{-i\vartheta}\right]^{1/(m+1)}$$

$(\alpha > 1)$. Then

$$\begin{aligned}\tilde{\varphi}(\zeta) &= \varphi(a\,\zeta)\\ &= \mathrm{Re}\,(A\,(a\,\zeta)^{m+1}) + o(|a\,\zeta|^{m+1})\\ &= \mathrm{Re}\,(\alpha\,\zeta^{m+1}) + o(|\zeta|^{m+1}).\end{aligned}$$

Now, if $z = \rho e^{i\theta}$, then

$$\varphi(\rho e^{i\theta}) = A\,\rho^{m+1}\cos((m+1)\,\theta) + o(\rho^{m+1}),$$

in particular, for $k = 0, 1, \ldots, m+1$:

(8.12)
$$\begin{aligned}\varphi(\rho e^{i\pi k/(m+1)}) &= A\,\rho^{m+1}\cos(\pi k) + o(\rho^{m+1})\\ &= (-1)^k A\,\rho^{m+1} + o(\rho^{m+1}).\end{aligned}$$

If R is sufficiently small, then for $z \in B_R$, i.e., $\rho < R$, it follows that the sign of the l. h. s. of (8.12) is $(-1)^k$. Hence if z traverses a Jordan curve C in $B_R(0)\backslash\{0\}$ (i.e., the index of C with respect to 0 is $\neq 0$), then $\varphi = \mathrm{Re}\,\Phi$ has to change sign at least $2m+2$ times.

Since Φ is a homeomorphism, $C = \Phi^{-1}\{|w| = \delta\} \subset B_\epsilon(0)$ is such a Jordan curve of index 1 with respect to 0 if $\delta > 0$ is sufficiently small. Hence, as z traverses C, $u = \mathrm{Re}\,w = \mathrm{Re}\,\Phi(z)$ changes sign exactly twice, and therefore $m = 0$ as required. □

Theorem 8.2.2. *Let* $w(z) = u + i\,v \in C^1(\Omega)$ *be a homeomorphic solution of* (8.1) *such that the Assumption* (A8.1) *is satisfied. Then*

(8.13)
$$Jw(z) = u_x v_y - u_y v_x \neq 0 \quad \text{in} \quad \Omega.$$

<u>Proof:</u> Suppose that

$$Jw(z_0) = 0$$

for some $z_0 \in \Omega$, and assume that $z_0 = 0$, $w(z_0) = 0$. Consider the linear system

$$w_x(0) \cdot \zeta = u_x(0)\,\xi + v_x(0)\,\eta = 0\,,$$

$$w_y(0) \cdot \zeta = u_y(0)\,\xi + v_y(0)\,\eta = 0\,.$$

There exists a solution $\zeta = (\xi, \eta)$, $|\zeta| = 1$. By Proposition 8.1.2, there exists a disc $B_\delta = B_\delta(0)$ in the w–plane and a function $\phi \in C^2(B_\delta)$ with $\phi(0) = 0$, $\phi_u(0) = \xi$, $\phi_v(0) = \eta$ such that $\varphi(z) = \phi(w(z))$ satisfies the differential inequality

$$|L\varphi| \le C(|D\varphi| + |\varphi|)$$

for $z \in B_R(0)$, $R = R(\delta)$. Now $\varphi(0) = \phi(w(0)) = 0$ and

$$\varphi_x(0) = \xi u_x(0) + \eta v_x(0) = 0\,,$$

$$\varphi_y(0) = \xi u_y(0) + \eta v_y(0) = 0\,.$$

Hence

$$\varphi = o(|z|)\,.$$

Corollary 7.4.2 implies $\varphi(z) \equiv 0$ or

$$\varphi(z) = \mathrm{Re}(A\,z^{m+1}) + o(|z|^{m+1}) \tag{8.14}$$

as $|z| \longrightarrow 0$ with $A \neq 0$ and $m \ge 1$.

Now consider the mapping

$$\Phi(z) = \varphi(z) + i\,(v(z)\,\xi - u(z)\,\eta)\,,$$

and remember that

$$\varphi(z) = u\,\xi + v\,\eta - g(v\,\xi - u\,\eta)\,.$$

Then Φ is univalent because $\Phi(z') = \Phi(z'')$ implies

$$(v(z') - v(z''))\,\xi - (u(z') - u(z''))\,\eta = 0\,,$$

$$(u(z') - u(z''))\,\xi + (v(z') - v(z''))\,\eta = 0\,,$$

from which $w(z') = w(z")$ and hence $z' = z"$ since w is univalent. This implies $\varphi = \mathrm{Re}\,\Phi \not\equiv 0$ and $m = 0$ in (8.14) by Lemma 8.2.1, a contradiction. □

8.3. Estimates for the Jacobian from below

Proposition 8.3.1. *Let* $\{w^{(k)}(z)\}_{k=1}^{\infty}$ *be a sequence of* C^1*-solutions of* (8.1) *such that the Assumption* (A8.1) *holds. Suppose that*

$$w^{(k)}(z) \longrightarrow w(z), \qquad Dw^{(k)}(z) \longrightarrow Dw(z) \tag{8.15}$$

uniformly in Ω *as* $k \longrightarrow \infty$. *Suppose that* $Jw^{(k)}(z) \neq 0$ *for* $z \in \Omega$. *Then either*

$$Jw(z) \neq 0 \;\; in \;\; \Omega \qquad or \qquad Jw(z) \equiv 0 \;\; in \;\; \Omega. \tag{8.16}$$

Proof: Assume that $Jw(z_0) = 0$, suppose that $z_0 = 0$, $w(z_0) = 0$. Then there is a solution $\zeta = (\xi, \eta)$, $|\zeta| = 1$ of the system

$$u_x(0)\,\xi + v_x(0)\,\eta = 0\,,$$

$$u_y(0)\,\xi + v_y(0)\,\eta = 0\,.$$

By Proposition 8.1.2, there exists a disc $B_\delta = B_\delta(0)$ and a function $\Phi \in C^2(B_\delta)$ such that $\varphi^{(k)}(z) = \Phi(w^{(k)}(z))$ satisfies

$$|L\varphi^{(k)}| \le C(|D\varphi^{(k)}| + |\varphi^{(k)}|)$$

in B_R, $R = R(\delta)$, because we may w.l.o.g. assume that $w^{(k)}(0) = 0$ and $|Dw^{(k)}| \le K$. Since $Jw^{(k)}(z) \neq 0$ in Ω, it follows that

$$D\varphi^{(k)} = \begin{bmatrix} u_x^{(k)} & v_x^{(k)} \\ u_y^{(k)} & v_y^{(k)} \end{bmatrix} \begin{bmatrix} \xi \\ \eta \end{bmatrix} \neq 0\,.$$

Furthermore

$$\varphi^{(k)} \longrightarrow \varphi, \quad D\varphi^{(k)} \longrightarrow D\varphi$$

uniformly in Ω, and $\varphi(0) = 0$, $D\varphi(0) = 0$. Proposition 7.4.3 yields $\varphi(z) \equiv 0$.
Hence, using the fact that $\varphi = u\,\xi + v\,\eta - g(v\,\xi - u\,\eta)$,

$$u_x\,\xi + v_x\,\eta - g'(v\,\xi - u\,\eta)(v_x\,\xi - u_x\,\eta) = 0\,,$$

$$u_y\,\xi + v_y\,\eta - g'(\;\ldots\;)(v_y\,\xi - u_y\,\eta) = 0\,,$$

from which

$$(u_x - g' v_x)\xi + (v_x + g' u_x)\eta = 0,$$

$$(u_y - g' v_y)\xi + (v_y + g' u_y)\eta = 0.$$

Therefore

$$0 = \det[\ldots] = (1 + (g')^2)(u_x v_y - u_y v_x)$$

as required. □

Theorem 8.3.2. *Let* $w(z)$ *be a homeomorphism from the closed unit disc* $\bar{B}$ *onto itself of class* $W^{1,2}$ *solving the system* (8.1). *Suppose that the Assumptions* (A8.1) *are satisfied, and assume that* $w(0) = 0$ *and*

$$\int_\Omega |Dw|^2 \, dx\, dy \leq M. \tag{8.17}$$

Then $w \in C^{1,\mu}_{loc}(B)$ *and the Jacobian* $Jw(z)$ *does not vanish in* B. *The following estimates hold for all* B_R, $0 < R < 1$:

$$\|w\|_{C^{1,\mu}(B_R)} \leq C, \tag{8.18}$$

$$Jw(z) \geq c > 0, \tag{8.19}$$

where the constants C *and* c *depend only on* μ, λ, Λ, H, M_0, M_1, M *and* R.

Proof: The $C^{1,\mu}$–estimate (8.18) is identical with the estimate (2.43) from Theorem 2.4.4. The non–vanishing of the Jacobian has been shown in Theorem 8.2.2.

To show the estimate (8.19), consider a disc B_R, $0 < R < 1$, and suppose that there is no estimate of the form

$$Jw(z) \geq c > 0 \quad \text{for} \quad z \in B_R.$$

Then there exists a sequence of points $\{z_k\}_{k=1}^\infty$ in B_R and a sequence of functions $\{w^{(k)}(z)\}_{k=1}^\infty$ satisfying the assumptions of the theorem such that

$$\lim_{k\to\infty} Jw^{(k)}(z_k) = 0.$$

By passing to a subsequence, we may assume that

$$z_k \longrightarrow z_0 \in \bar{B}_R,$$

$$w^{(k)}(z) \longrightarrow w(z), \quad Dw^{(k)}(z) \longrightarrow Dw(z)$$

uniformly in each compact part of B. Hence

$$Jw(z_0) = 0.$$

Now Theorem 8.2.2 implies that

$$Jw^{(k)}(z) \neq 0 \quad \text{in} \quad \Omega,$$

and therefore

$$\text{(8.20)} \qquad Jw(z) \equiv 0$$

by Proposition 8.3.1.

This is impossible however, because the conditions $|w^{(k)}(z)| = 1$ for $|z| = 1$ and $w^{(k)}(0) = 0$ yield the equicontinuity of $\{w^{(k)}(z)\}_{k=1}^{\infty}$ in $\bar{B}$ by the Courant–Lebesgue lemma, Lemma 1.6.3. Hence there exists a R, $0 < R < 1$, such that

$$|w^{(k)}(z)| > \tfrac{1}{2} \quad \text{for} \quad |z| \geq R,$$

from which

$$B_{1/2} \subset w^{(k)}(B_R)$$

for all $k \in \mathbb{N}$, and therefore

$$\begin{aligned}\int_{B_R} |Jw^{(k)}(z)|\, dx\, dy &= \int_{w^{(k)}(B_R)} du\, dv \\ &\geq \int_{B_{1/2}} du\, dv \\ &= \tfrac{\pi}{4}.\end{aligned}$$

By passing to the limit,

$$\int_{B_R} |Jw(z)|\, dx\, dy \geq \tfrac{\pi}{4},$$

which contradicts (8.20). □

Chapter 9. A PRIORI ESTIMATES FOR MONGE–AMPÈRE EQUATIONS

9.1. Characteristic parameters for nonlinear elliptic equations

Let $z = z(x,y) \in C^2(\Omega)$ be a solution of the elliptic equation

$$F(x,y,z,p,q,r,s,t) = 0, \tag{9.1}$$

i.e., by Definition 3.1.1, the characteristic form is positive definite,

$$F_r \xi^2 + F_s \xi\eta + F_t \eta^2 > 0 \tag{9.2}$$

for all $(\xi,\eta) \in \mathbb{R}^2$, $(x,y) \in \Omega$, and $F_r > 0$. Here $F_r = F_r(x,y,z(x,y),\dots,t(x,y)),\dots,$ $F_t = F_t(\dots)$. The function F is continuous and differentiable with respect to r, s, t. One can use Theorem 6.3.2, resp. Remark 6.4.1, to introduce new independent variables $w = (u,v)$ in the large such that

$$ds^2 = \Lambda\,(du^2 + dv^2), \qquad \Lambda \neq 0, \tag{9.3}$$

where one takes

$$\begin{aligned} ds^2 &= F_t\,dx^2 - F_s\,dx\,dy + F_r\,dy^2 \\ &= a(x,y)\,dx^2 + 2\,b(x,y)\,dx\,dy + c(x,y)\,dy^2 \end{aligned} \tag{9.4}$$

as the characteristic form. The regularity requirements of Remark 6.4.1, namely that $a, b, c \in C^\mu(\Omega)$ for some μ, $0 < \mu < 1$, are e.g. satisfied if $z \in C^{2,\mu}(\Omega)$ and $F_r, F_s, F_t \in C^\mu$ with respect to all variables and if F_r, F_s, F_t are Lipschitz continuous with respect to r, s, t. The mapping $(u,v) = (u(x,y), v(x,y))$ is then a diffeomorphism.

Definition 9.1.1. The parameters u and v are called *characteristic* with respect to the equation $F = 0$.

9.2. Characteristic parameters for Monge–Ampère equations

Consider the Monge–Ampère equation

$$A\,r + 2B\,s + C\,t + (r\,t - s^2) = E \tag{9.5}$$

or equivalently,

$$(r + C)\,(t + A) - (s - B)^2 = \Delta, \tag{9.6}$$

where

$$\Delta = AC - B^2 + E > 0, \qquad t + A > 0, \tag{9.7}$$

is the ellipticity condition. The coefficients A, B, C, E depend on the variables x, y, z, p, q. According to the regularity theory, Theorem 3.4.3, solutions of class $C^{1,1}(\Omega)$ belong in fact to the class $C^{2,\mu}(\Omega)$ if $A, B, C, E \in C^{\mu}(\Omega \times \mathbb{R}^3)$ for some μ, $0 < \mu < 1$.

Proposition 9.2.1. *Suppose that* $z \in C^{1,1}(\Omega)$ *is a solution of the elliptic Monge–Ampère equation* (9.5). *Furthermore let* $A, B, C \in C^1(\Omega \times \mathbb{R}^3)$, $\Delta \in C^{\mu}(\Omega \times \mathbb{R}^3)$ *for some* μ, $0 < \mu < 1$, *and let* $\bar{B}_R = \bar{B}_R(x_0, y_0) \subset \Omega$. *Then there exists a homeomorphism* $(x,y) = (x(u,v), y(u,v))$ *from* $\bar{B} = \{u^2 + v^2 \leq 1\}$ *onto* $\bar{B}_R$ *of class* $C^{1,\mu}_{loc}(B)$ *with* $x(0) = x_0$, $y(0) = y_0$ *and*

$$J(x,y) = x_u y_v - x_v y_u \neq 0 \tag{9.8}$$

such that

$$\frac{t + A}{\sqrt{\Delta}} = \frac{|Dx|^2}{J(x,y)}, \tag{9.10}$$

$$-\frac{s - B}{\sqrt{\Delta}} = \frac{Dx \cdot Dy}{J(x,y)}, \tag{9.11}$$

$$\frac{r + C}{\sqrt{\Delta}} = \frac{|Dy|^2}{J(x,y)}. \tag{9.12}$$

Furthermore

$$Lx = h_1 |Dx|^2 + h_2 Dx \cdot Dy + h_3 |Dy|^2 + h_4 Dx \wedge Dy, \tag{9.13}$$

$$Ly = \tilde{h}_1 |Dx|^2 + \tilde{h}_2 Dx \cdot Dy + \tilde{h}_3 |Dy|^2 + \tilde{h}_4 Dx \wedge Dy, \tag{9.14}$$

where

$$L = -\frac{1}{\sqrt{\Delta}} \left[\frac{\partial}{\partial u}\left[\sqrt{\Delta}\frac{\partial}{\partial u}\right] + \frac{\partial}{\partial v}\left[\sqrt{\Delta}\frac{\partial}{\partial v}\right]\right], \tag{9.15}$$

and

$$\begin{aligned} h_1(x,y) &= -B_q, \\ h_2(x,y) &= A_q + B_p, \\ h_3(x,y) &= -A_p, \\ h_4(x,y) &= -\frac{1}{\sqrt{\Delta}}\{A_x + A_z p + B_y + B_z q - A_p C + (A_q + B_p) B - B_q A\}, \end{aligned} \tag{9.16}$$

$$\begin{aligned} \tilde{h}_1(x,y) &= -C_q, \\ \tilde{h}_2(x,y) &= B_q + C_p, \\ \tilde{h}_3(x,y) &= -B_p, \\ \tilde{h}_4(x,y) &= \frac{1}{\sqrt{\Delta}}\{B_x + B_z p + C_y + C_z q - B_p C + (B_q + C_p) B - C_q A\}. \end{aligned} \tag{9.17}$$

Proof: The characteristic form associated with the Monge–Ampère equation (9.5) is

$$\begin{aligned} ds^2 &= F_t\,dx^2 - F_s\,dx\,dy + F_r\,dy^2 \\ &= (r+C)\,dx^2 + 2(s-B)\,dx\,dy + (t+A)\,dy^2 \\ &= (r+\tilde{A})\,dx^2 + 2(s+\tilde{B})\,dx\,dy + (t+\tilde{C})\,dy^2 . \end{aligned}$$

Theorem 6.3.2 yields the existence of the characteristic parameters (u,v) together with the conformality relations (9.10, 11, 12). Furthermore, according to (6.52, 53),

$$-\sqrt{\Delta}\,Lx = L_\Delta x = (\tilde{C}_x - \tilde{B}_y)(x_u y_v - x_v y_u) = (A_x + B_y)\,J(x,y),$$

$$-\sqrt{\Delta}\,Ly = L_\Delta y = (\tilde{A}_y - \tilde{B}_x)(x_u y_v - x_v y_u) = (B_x + C_y)\,J(x,y).$$

We compute, using the conformality relations (9.10, 11, 12):

$$\begin{aligned} A_x + B_y &= \frac{\partial}{\partial x}(A(x,y,z(x,y),\ldots,q(x,y))) + \frac{\partial}{\partial y}(B(\ldots)) \\ &= A_x + A_z p + A_p r + A_q s + B_y + B_z q + B_p s + B_q t \\ &= A_x + A_z p + B_y + B_z q - A_p C + (A_q + B_p) B - B_q A \\ &\qquad + A_p (r+C) + (A_q + B_p)(s-B) + B_q (t+A) \\ &= -\sqrt{\Delta}\,h_4 - h_3 (r+C) + h_2 (s-B) - h_1 (t+A) \\ &= -\sqrt{\Delta}\left[h_4 + h_3 \frac{|Dy|^2}{J(x,y)} + h_2 \frac{Dx\cdot Dy}{J(x,y)} + h_1 \frac{|Dx|^2}{J(x,y)}\right] \\ &= -\frac{\sqrt{\Delta}}{J(x,y)}\left(h_1 |Dx|^2 + h_2\, Dx\cdot Dy + h_3 |Dy|^2 + h_4\, Dx\wedge Dy\right), \end{aligned}$$

and, similarly,

$$\begin{aligned} B_x + C_y &= B_x + B_z p + B_p r + B_q s + C_y + C_z q + C_p s + C_q t \\ &= B_x + B_z p + C_y + C_z q - B_p C + (B_q + C_p) B - C_q A \\ &\qquad + B_p (r+C) + (B_q + C_p)(s-B) + C_q (t+A) \\ &= -\sqrt{\Delta}\,\tilde{h}_4 - \tilde{h}_3 (r+C) + \tilde{h}_2 (s-B) - \tilde{h}_1 (t+A) \\ &= -\frac{\sqrt{\Delta}}{J(x,y)}\left(\tilde{h}_1 |Dx|^2 + \tilde{h}_2\, Dx\cdot Dy + \tilde{h}_3 |Dy|^2 + \tilde{h}_4\, Dx\wedge Dy\right). \quad \square \end{aligned}$$

9.3. A priori estimates

Lemma 9.3.1. *Let* $z(x,y) \in C^{1,1}(\Omega)$ *be a solution of the Monge–Ampère equation* (9.5). *Suppose that*

(9.18) $$|A|,\dots,|E| \le a, \qquad \Delta \ge \frac{1}{c}$$

and

(9.19) $$|Dz| \le K_1.$$

Then the mapping $(x,y) = (x(u,v), y(u,v))$, $(u,v) \in B = \{u^2+v^2<1\}$ *from Proposition* 9.2.1 *satisfies the estimate*

(9.20) $$\int_B (|Dx|^2 + |Dy|^2)\,du\,dv \le C(a,c,K_1)\,R.$$

Proof: By the conformality relations (9.10, 11, 12),

$$\begin{aligned}\int_B (|Dx|^2 + |Dy|^2)\,du\,dv &= \int_{B_R(x_0,y_0)} \frac{t+A+r+C}{\sqrt{\Delta}}\, J(x,y)\,\frac{dx\,dy}{J(x,y)}\\ &\le \sqrt{c}\int_{B_R}(A+C+r+t)\,dx\,dy\\ &\le CR^2 + \int_{\partial B_R} Dz\cdot\nu\,d\sigma\\ &\le C(R^2+R) \le CR\end{aligned}$$

if $R \le 1$. Here we employed a general divergence theorem. □

Assumption (A9.1). Suppose that $z \in C^{1,1}(\Omega)$ satisfies the estimate

(9.21) $$|Dz| \le K_1.$$

Assumption (A9.1'). Suppose that

(9.22) $$|Dz| + [Dz]_\sigma^\Omega \le K_{1,\sigma}$$

for some σ, $0<\sigma<1$.

Assumption (A9.2). (i) Let $A, B, C \in C^1(\Omega\times\mathbb{R}^3)$, $\Delta \in C^\mu(\Omega\times\mathbb{R}^3)$ for some μ, $0<\mu<1$, such that

(9.23) $$|A|,\ |B|,\ |C|,\ |E| \le a,$$

(9.24) $$|A_x|,\dots,|C_q| \le b,$$

(9.25) $$\Delta = AC - B^2 + E \ge \frac{1}{c},$$

(9.26) $$[\Delta]_\mu^{\Omega\times\mathbb{R}^3} \le H.$$

(ii) Suppose in addition that the functions

$$\begin{aligned}\phi_1(x,y) &= A_p,\\ \phi_2(x,y) &= A_q + 2B_p,\\ \phi_3(x,y) &= C_p + 2B_q,\\ \phi_4(x,y) &= C_q\end{aligned} \tag{9.27}$$

are Lipschitz continuous with

$$[\phi_1]^{\Omega}_{0,1}, \dots, [\phi_4]^{\Omega}_{0,1} \le d. \tag{9.28}$$

Assumption (A9.3). Suppose that Δ is a function of the variables x, y, z only.

Theorem 9.3.2. *Let* $z \in C^{1,1}(\Omega)$ *be a solution of the Monge–Ampère equation* (9.5). *Suppose that the assumptions* (A9.1, 2, 3) *or* (A9.1′, 2) *are satisfied. Then* $z \in C^{2,\mu}_{loc}(\Omega)$, *and for each* $\Omega' \subset\subset \Omega$ *there is an estimate of the form*

$$||D^2z||_{C^{\mu}(\Omega')} \le C. \tag{9.29}$$

In the case of the assumptions (A9.1, 2, 3), *the constant* C *depends only on* μ, a, b, c, H, d, K_1, *and* $\mathrm{dist}(\Omega', \partial\Omega)$. *If the conditions* (A9.1′, 2) *are imposed, then the constant* C *depends in addition to these constants also on* σ *and* $K_{1,\sigma}$.

Proof: Let the assumptions (A9.1, 2, 3) be imposed. The function

$$\Delta(x,y) = \Delta(x,y,z(x,y))$$

is then of class $C^{\mu}(\Omega)$ and an estimate of the form

$$[\Delta]^{\Omega}_{\mu} \le C$$

holds true. Therefore, by Theorem 3.4.3, $z \in C^{2,\mu}_{loc}(\Omega)$ together with the estimate

$$[D^2z]^{\Omega'}_{\mu} \le C(\dots, \mathrm{dist}(\Omega', \partial\Omega)). \tag{9.30}$$

To estimate the absolute value of D^2z, consider the homeomorphism $(x,y) = (x(u,v), y(u,v))$ from $\bar{B}$ onto $\bar{B}_R$, $\bar{B}_R = \bar{B}_R(x_0, y_0) \subset \Omega$, from Proposition 9.2.1. (x,y) is of class $C^{1,\mu}_{loc}(B)$ with finite Dirichlet integral (by Lemma 9.3.1) and satisfies the relations

$$\text{(9.31)} \qquad \frac{t+A}{\sqrt{\Delta}} = \frac{|Dx|^2}{J(x,y)},$$

$$\text{(9.32)} \qquad -\frac{s-B}{\sqrt{\Delta}} = \frac{Dx \cdot Dy}{J(x,y)},$$

$$\text{(9.33)} \qquad \frac{r+C}{\sqrt{\Delta}} = \frac{|Dy|^2}{J(x,y)}.$$

Furthermore, (x,y) solves a system of the form

$$Lx = h_1 |Dx|^2 + h_2 Dx \cdot Dy + h_3 |Dy|^2 + h_4 Dx \wedge Dy,$$

$$Ly = \tilde{h}_1 |Dx|^2 + \tilde{h}_2 Dx \cdot Dy + \tilde{h}_3 |Dy|^2 + \tilde{h}_4 Dx \wedge Dy$$

with

$$L = -\frac{1}{\sqrt{\Delta}} \left[\frac{\partial}{\partial u} \left[\sqrt{\Delta} \frac{\partial}{\partial u} \right] + \frac{\partial}{\partial v} \left[\sqrt{\Delta} \frac{\partial}{\partial v} \right] \right].$$

The function $h_1, \ldots, \tilde{h}_4$ are defined by (9.16,17).

In order to apply Theorem 8.3.2, we assume w.l.o.g. that $B_R(x_0, y_0) = B = \{x^2 + y^2 < 1\}$. This is possible because the function $z(x_0 + Rx, y_0 + Ry)$ solves a Monge–Ampère equation of the type (9.5) with coefficients of the form $R^2 A(x_0 + Rx, y_0 + Ry, \ldots)$, which satisfy Assumption (A9.2) with constants which now depend also on R.

Note that the role of the variables x, y and u, v is reversed in Theorem 8.3.2. The assumption of Theorem 8.3.2 can be verified, in particular the Lipschitz continuity of

$$\omega_1(x,y) = \tilde{h}_1(x,y) = -C_q = -\phi_4,$$

$$\omega_2(x,y) = h_1(x,y) - \tilde{h}_2(x,y) = -2B_q - C_p = -\phi_3,$$

$$\omega_3(x,y) = h_2(x,y) - \tilde{h}_3(x,y) = A_q + 2B_p = \phi_2,$$

$$\omega_4(x,y) = h_3(x,y) = -A_p = -\phi_1.$$

The following estimates are therefore true in any disc $B_\rho = \{u^2 + v^2 < \rho^2\}$, $0 < \rho < 1$:

$$\|x,y\|_{C^{1,\mu}(B_\rho)} \le C(\ldots,\rho),$$

$$J(x,y) \ge c(\ldots,\rho) > 0,$$

in particular, by taking $\rho = 1/2$,

$$|Dx(0)|, \quad |Dy(0)| \le C,$$

$$(J(x,y))(0) \ge c.$$

The relations (9.31, 32, 33) translate this into the estimate

$$|D^2 z(0)| \le C,$$

which actually means that

$$|D^2 z(x_0, y_0)| \le C(\dots, R) \tag{9.34}$$

since we assumed that $B_R(x_0, y_0) = \{x^2 + y^2 < 1\}$. The estimates (9.30, 34) are identical with the statement.

Some complications arise when the conditions (A9.1′, 2) are imposed: The function

$$\Delta(x,y) = \Delta(x, y, z(x,y)), \dots, q(x,y))$$

belongs then to the Hölder class $C^{\mu}(\Omega)$, but there is only an estimate of the form

$$[\Delta]^{\Omega}_{\mu'} \le C(\dots, \sigma, K_{1,\sigma})$$

with

$$\mu' = \sigma\mu.$$

Estimates for the absolute value of $D^2 z$ can then be derived as above. This means that the Assumption (A9.1′) is true for $\sigma = 1$, and the Hölder estimate (9.30) follows again from Theorem 3.4.3. □

<u>Remark 9.3.3.</u> The Assumption (A9.1′) can be verified for the simple Monge–Ampère equation

$$r t - s^2 = f(x, y, z, p, q) > 0. \tag{9.35}$$

This Hölder estimate for Dz can be found in Heinz [H 5]. Theorem 9.3.2 provides thus sharp second derivative estimates for (9.35) without invoking the full thrust of [H 5]. This is in contrast to Schulz [SZ 2], where the Heinz estimates of [H 5] are employed to yield dilation estimates for the Legendre like transformation

$$\begin{aligned} u &= x, \\ v &= q(x,y). \end{aligned} \tag{9.36}$$

These estimates represent quantitative versions of the Alexandrov theorem, that if the generalized Gauß curvature of a convex surface is pinched between two positive numbers, then the surface does not contain straight line segments.

9.4. Estimates for a general class of Monge–Ampère equations

Consider the Monge–Ampère equation

(9.37) $$(r+C)(t+A)-(s-B)^2=\Delta>0$$

subject to the following condition:

Assumption (A9.4). Suppose that

(9.38) $$\Delta = K(x,y,z)\,D(x,y,z,p,q),$$

where $K \in C^{\mu}(\Omega \times \mathbb{R})$ for some μ, $0<\mu<1$, and $D \in C^1(\Omega \times \mathbb{R}^3)$ such that

(9.39) $$|K|,\ |D| \le a,$$

(9.40) $$|D_x|,\dots,|D_q| \le b,$$

(9.41) $$K,\ D \ge \frac{1}{c},$$

(9.42) $$[K]_{\mu}^{\Omega\times\mathbb{R}} \le H.$$

Theorem 9.4.1. *Suppose that* $z \in C^{1,1}(\Omega)$ *is a solution of* (9.37) *such that the assumptions* (A9.1, 2, 4) *are satisfied. Then* $z \in C^{2,\mu}_{loc}(\Omega)$, *and for each* $\Omega' \subset\subset \Omega$ *there is an estimate of the form*

(9.43) $$\|D^2z\|_{C^{\mu}(\Omega')} \le C(\mu, a, b, c, H, d, K_1, \operatorname{dist}(\Omega', \partial\Omega)).$$

The proof of Theorem 9.4.1 is along the lines of the proof of Theorem 9.3.2, invoking the following version of Proposition 9.2.1:

Proposition 9.4.2. *Let* $\bar{B}_R = \bar{B}_R(x_0, y_0) \subset \Omega$. *Then there is a homeomorphism* (x,y) *from* $\bar{B} = \{u^2+v^2 \le 1\}$ *onto* $\bar{B}_R$ *of class* $C^{1,\mu}_{loc}(B)$ *with* $x(0)=x_0$, $y(0)=y_0$ *and* $J(x,y) \ne 0$ *such that*

(9.44) $$\frac{t+A}{\sqrt{\Delta}} = \frac{|Dx|^2}{J(x,y)},$$

(9.45) $$-\frac{s-B}{\sqrt{\Delta}} = \frac{Dx \cdot Dy}{J(x,y)},$$

(9.46) $$\frac{r+C}{\sqrt{\Delta}} = \frac{|Dy|^2}{J(x,y)}.$$

Furthermore,

$$(9.47)\qquad Lx = h_1\,|Dx|^2 + h_2\,Dx\cdot Dy + h_3\,|Dy|^2 + h_4\,Dx\wedge Dy,$$

$$(9.48)\qquad Ly = \tilde h_1\,|Dx|^2 + \tilde h_2\,Dx\cdot Dy + \tilde h_3\,|Dy|^2 + \tilde h_4\,Dx\wedge Dy,$$

where

$$(9.49)\qquad L = -\frac{1}{\sqrt{K}}\left[\frac{\partial}{\partial u}\left[\sqrt{K}\frac{\partial}{\partial u}\right] + \frac{\partial}{\partial v}\left[\sqrt{K}\frac{\partial}{\partial v}\right]\right].$$

and

$$(9.50)\qquad \begin{aligned} h_1(x,y) &= -B_q + \tfrac{1}{2D}(D_x + D_z p - D_p C + D_q B),\\ h_2(x,y) &= A_q + B_p + \tfrac{1}{2D}(D_y + D_z q + D_p B - D_q A),\\ h_3(x,y) &= -A_p,\\ h_4(x,y) &= -\frac{1}{\sqrt{\Delta}}\left[A_x + A_z p + B_y + B_z q - A_p C + (A_q + B_p)B - B_q A - \frac{D_p K}{2}\right]. \end{aligned}$$

$$(9.51)\qquad \begin{aligned} \tilde h_1(x,y) &= -C_q,\\ \tilde h_2(x,y) &= B_q + C_p + \tfrac{1}{2D}(D_x + D_z p - D_p C + D_q B),\\ \tilde h_3(x,y) &= -B_p + \tfrac{1}{2D}(D_y + D_z q + D_p B - D_q A),\\ \tilde h_4(x,y) &= \frac{1}{\sqrt{\Delta}}\left[B_x + B_z p + C_y + C_z q - B_p C + (B_q + C_p)B - C_q A - \frac{D_q K}{2}\right]. \end{aligned}$$

<u>**Proof:**</u> Consider the characteristic form

$$\begin{aligned} ds^2 &= (r + C)\,dx^2 + 2(s - B)\,dx\,dy + (t + A)\,dy^2\\ &= a\,dx^2 + 2b\,dx\,dy + c\,dy^2. \end{aligned}$$

As in Remark 6.1.7, consider the corresponding Beltrami systems (6.30, 31) and (6.32, 33), i.e.,

$$\sqrt{K}x_u = \frac{bx_v + cy_v}{\sqrt{D}},$$

$$\sqrt{K}x_v = \frac{-bx_u - cy_u}{\sqrt{D}}$$

and

$$\sqrt{K}y_u = \frac{-ax_v - by_v}{\sqrt{D}},$$

$$\sqrt{K}y_v = \frac{ax_u - by_u}{\sqrt{D}}.$$

If $ds^2 \in C^1$, then by differentiation,

$$L_K x = \left[\left[\frac{c}{\sqrt{D}}\right]_x - \left[\frac{b}{\sqrt{D}}\right]_y\right](x_u y_v - x_v y_u),$$

$$L_K y = \left[\left[\frac{a}{\sqrt{D}}\right]_y - \left[\frac{b}{\sqrt{D}}\right]_x\right](x_u y_v - x_v y_u),$$

here

$$L_K = \frac{\partial}{\partial u}\left[\sqrt{K}\frac{\partial}{\partial u}\right] + \frac{\partial}{\partial v}\left[\sqrt{K}\frac{\partial}{\partial v}\right].$$

Hence as in Remark 6.1.8,

$$L_K x = \left[\frac{A_x}{\sqrt{D}} + \frac{B_y}{\sqrt{D}} + (t+A)\left[\frac{1}{\sqrt{D}}\right]_x - (s-B)\left[\frac{1}{\sqrt{D}}\right]_y\right](x_u y_v - x_v y_u)$$

$$(9.52) \qquad = \frac{1}{\sqrt{D}}\left[A_x + B_y - (t+A)\frac{D_x}{2D} + (s-B)\frac{D_y}{2D}\right](x_u y_v - x_v y_u),$$

$$L_K y = \left[\frac{B_x}{\sqrt{D}} + \frac{C_y}{\sqrt{D}} + (s-B)\left[\frac{1}{\sqrt{D}}\right]_x + (r+C)\left[\frac{1}{\sqrt{D}}\right]_y\right](x_u y_v - x_v y_u)$$

$$(9.53) \qquad = \frac{1}{\sqrt{D}}\left[B_x + C_y + (s-B)\frac{D_x}{2D} - (r+C)\frac{D_y}{2D}\right](x_u y_v - x_v y_u).$$

The requirement $z \in C^3$, which was used in the derivation of the system (9.52, 53), can be removed by approximation, i.e., a theorem analog to Theorem 6.3.2 can be proven with the system (6.52, 53) replaced by (9.52, 53) if $z \in C^{2,\mu}$.

As in the proof of Proposition 9.2.1 one computes, using the conformality relations (9.44,45,46):

$$\begin{aligned}
A_x + B_y - (t+A)\frac{D_x}{2D} + (s-B)\frac{D_y}{2D} &= \frac{\partial}{\partial x}(A(x,y,z(x,y),\dots)) + \frac{\partial}{\partial y}(B(\dots)) - \dots \\
&= A_x + A_z p + B_y + B_z q - A_p C + (A_q + B_p)B - B_q A \\
&\quad + A_p(r+C) + (A_q+B_p)(s-B) + B_q(t+A) \\
&\quad - \frac{t+A}{2D}(D_x + D_z p - D_p C + D_q B + D_p(r+C) + D_q(s-B)) \\
&\quad + \frac{s-B}{2D}(D_y + D_z q + D_p B - D_q A + D_p(s-B) + D_q(t+A)) \\
&= A_x + A_z p + B_y + B_z q - A_p C + (A_q + B_p)B - B_q A \\
&\quad + (t+A)(B_q - \frac{1}{2D}(D_x + D_z p - D_p C + D_q B)) \\
&\quad + (s-B)(A_q + B_p + \frac{1}{2D}(D_y + D_z q + D_p B - D_q A)) \\
&\quad + (r+C)A_p - \frac{D_p}{2D}((t+A)(r+C) - (s-B)^2) \\
&= -\frac{\sqrt{\Delta}}{J(x,y)}(h_1|Dx|^2 + h_2 Dx\cdot Dy + h_3|Dy|^2 + h_4 Dx\wedge Dy).
\end{aligned}$$

Similarly,

$$\begin{aligned}
B_x + C_y + (s-B)\frac{D_x}{2D} - (r+C)\frac{D_y}{2D} \\
&= B_x + B_z p + C_y + C_z q - B_p C + (B_q + C_p) B - C_q A \\
&\quad + B_p (r+C) + (B_q + C_p)(s-B) + C_q (t+A) \\
&\quad + \frac{s-B}{2D}(D_x + D_z p - D_p C + D_q B + D_p (r+C) + D_q (s-B)) \\
&\quad - \frac{r+C}{2D}(D_y + D_z q + D_p B - D_q A + D_p (s-B) + D_q (t+A)) \\
&= B_x + B_z p + C_y + C_z q - B_p C + (B_q + C_p) B - C_q A + (t+A) C_q \\
&\quad + (s-B)(B_q + C_p + \frac{1}{2D}(D_x + D_z p - D_p C + D_q B)) \\
&\quad + (r+C)(B_p - \frac{1}{2D}(D_y + D_z q + D_p B - D_q A)) \\
&\quad - \frac{D_q}{2D}((r+C)(t+A) - (s-B)^2) \\
&= -\frac{\sqrt{\Delta}}{J(x,y)}(\tilde{h}_1 |Dx|^2 + \tilde{h}_2 Dx \cdot Dy + \tilde{h}_3 |Dy|^2 + \tilde{h}_4 Dx \wedge Dy). \quad \square
\end{aligned}$$

9.5. The Heinz–Lewy counterexample

The following counterexample shows, that structural conditions of some kind have to be imposed on the coefficients A, B, C, E of the Monge–Ampère equation (9.5), in order to have second derivative estimates for the solutions:

Example 9.5.1. Let $\epsilon > 0$ and consider the initial value problem

$$f'(x)^3 + \epsilon f'(x) = x, \qquad f(0) = 0. \tag{9.54}$$

There is a real analytic solution $f = f_\epsilon(x)$ such that

$$|f'(x)| \leq 1 \qquad \text{for} \qquad |x| \leq 1 \tag{9.55}$$

and

$$f''(x) = \frac{1}{3f'(x)^2 + \epsilon}. \tag{9.56}$$

The function

$$z = z_\epsilon(x,y) = f(x) + \frac{y^2}{2} \tag{9.57}$$

solves the Monge–Ampère equation

$$\begin{aligned} A r + (r t - s^2) &= (-1 + 3 p^2 + \epsilon) r + (r t - s^2) \\ &= (-1 + 3 f'(x)^2 + \epsilon) f''(x) + f''(x) \\ &= 1 \\ &= E . \end{aligned}$$

Furthermore, for $x^2 + y^2 \leq 1$:

(9.59) $$|p| = |f'(x)| \leq 1, \qquad |q| = |y| \leq 1$$

and

(9.60) $$|A|, \ldots, |E_{qq}| \leq 5,$$

(9.61) $$\Delta = 1 .$$

But

(9.62) $$r(0,0) = f''(0) = \frac{1}{3 f'(0)^2 + \epsilon} = \frac{1}{\epsilon} \longrightarrow +\infty$$

as $\epsilon \longrightarrow 0$.

Chapter 10. REGULARITY AND A PRIORI ESTIMATES FOR LOCALLY CONVEX SURFACES

10.1. Fundamental formulas in classical surface theory

Let Ω be a domain in the $(u^1,u^2) = (u,v)$–plane. A *regular surface* Σ is given by a map X of class $C^{1,1}(\Omega,\mathbb{R}^3)$ such that

$$D_1X \wedge D_2X \neq 0. \tag{10.1}$$

X is called the *radius vector.* The *unit normal* is the vector

$$\nu = \frac{D_1X\wedge D_2X}{|D_1X\wedge D_2X|}. \tag{10.2}$$

The *first fundamental form* is

$$\begin{aligned} I_\Sigma &= |dX|^2 \\ &= D_iX\cdot D_jX\,du^i du^j \\ &= g_{ij}\,du^i du^j \\ &= E\,du^2 + 2F\,du\,dv + G\,dv^2, \end{aligned} \tag{10.3}$$

and

$$g = \det I_\Sigma = |D_1X\wedge D_2X|^2 > 0 \tag{10.4}$$

by the Lagrange identity. The *second fundamental form*

$$\begin{aligned} II_\Sigma &= -dX\cdot d\nu \\ &= D_{ij}X\cdot \nu\,du^i du^j \\ &= h_{ij}\,du^i du^j \\ &= L\,du^2 + 2M\,du\,dv + N\,dv^2 \end{aligned} \tag{10.5}$$

is defined almost everywhere.

The *Gauß equations*

$$D_{ij}X = \Gamma^k_{ij}D_kX + h_{ij}\nu \tag{10.6}$$

motivate the definition of II_Σ to study curvature properties of Σ. The Γ^k_{ij}'s are the *Christoffel symbols* (of the second kind). One easily computes

(10.7) $$\Gamma^k_{ij} = \frac{1}{2} g^{k\ell}(D_j g_{i\ell} + D_i g_{j\ell} - D_\ell g_{ij}).$$

Here

(10.8) $$[g^{k\ell}] = [g_{ij}]^{-1} = \frac{1}{g}\begin{bmatrix} G & -F \\ -F & E \end{bmatrix}.$$

Note that

(10.9) $$D_k g_{ij} = \Gamma^\ell_{ik} g_{\ell j} + \Gamma^\ell_{jk} g_{\ell i},$$

and hence

(10.10) $$D_k g = 2g(\Gamma^1_{1k} + \Gamma^2_{2k}).$$

The *Weingarten equations* are

(10.11) $$D_i \nu = -h_{ij} g^{jk} D_k X.$$

The *Codazzi–Mainardi* equations can easily be derived as integrability conditions for the Gauß–Weingarten equations if $X \in C^{2,1}$:

(10.12) $$D_j h_{ik} - D_i h_{jk} = \Gamma^\ell_{jk} h_{i\ell} - \Gamma^\ell_{ik} h_{j\ell}.$$

The remaining integrability conditions are the *Gauß equations*

(10.13) $$D_\ell \Gamma^k_{ij} - D_j \Gamma^k_{i\ell} = \Gamma^m_{i\ell}\Gamma^k_{mj} - \Gamma^m_{ij}\Gamma^k_{m\ell} + (h_{ij}h_{\ell m} - h_{i\ell}h_{jm})\, g^{mk}.$$

The *theorema egregium* of Gauß can be obtained by letting $i = j = 1$, $k = \ell = 2$ in (10.13), namely that the *Gauß curvature*

(10.14) $$K = \frac{\det II_\Sigma}{\det I_\Sigma} = \frac{h}{g} = \frac{h_{11}h_{22} - h_{12}^2}{g_{11}g_{22} - g_{12}^2}$$

depends only on the coefficients of I_Σ and their first and second derivatives. Note that the definition (10.14) and the theorema egregium only requires the regularity $X \in C^{1,1}$ and $I_\Sigma \in C^{1,1}$.

Finally,

(10.15) $$H = \frac{h_{ij} g^{ij}}{2}$$

is the *mean curvature* and

(10.16) $$d\sigma = \sqrt{g}\, du^1\, du^2$$

is the *area element* of Σ.

10.2. The prescribed Gauß curvature equation

Suppose that Σ is a graph over the x, y–plane, i.e.,

$$X = (x, y, z(x, y)) \tag{10.17}$$

with $z \in C^{1,1}(\Omega)$. Then

$$D_1 X = (1, 0, z_x), \qquad D_2 X = (0, 1, z_y), \tag{10.18}$$

$$\nu = \frac{(-z_x, -z_y, 1)}{\sqrt{1 + z_x^2 + z_y^2}}, \tag{10.19}$$

$$D_{11} X = (0, 0, z_{xx}), \quad D_{12} X = (0, 0, z_{xy}), \quad D_{22} X = (0, 0, z_{yy}). \tag{10.20}$$

Hence

$$I_\Sigma = (1 + z_x^2)\, dx^2 + z_x z_y\, dx\, dy + (1 + z_y^2)\, dy^2, \tag{10.21}$$

$$II_\Sigma = \frac{1}{\sqrt{1 + z_x^2 + z_y^2}} (z_{xx}\, dx^2 + 2 z_{xy}\, dx\, dy + z_{yy}\, dy^2), \tag{10.22}$$

$$K = \frac{z_{xx} z_{yy} - z_{xy}^2}{(1 + z_x^2 + z_y^2)^2}. \tag{10.23}$$

The function $z = z(x, y)$ satisfies therefore the *prescribed Gauß curvature equation*

$$rt - s^2 = K(x, y, z)\,(1 + p^2 + q^2)^2 \tag{10.24}$$

which is elliptic if $K > 0$, i.e., if Σ is convex (or concave).

The results of Chapters 3 and 9 can be applied if $K \in C^\mu(\mathbb{R}^3)$ for some μ, $0 < \mu < 1$, to yield, according to Theorem 9.3.2:

Theorem 10.2.1. *Suppose that* Σ *is a graph over the* x, y*–plane with radius vector* $X \in C^{1,1}(\Omega, \mathbb{R}^3)$ *and with positive Gauß curvature* $K \in C^\mu(\mathbb{R}^3)$ *for some* μ, $0 < \mu < 1$. *Then* $X \in C^{2,\mu}_{loc}(\Omega, \mathbb{R}^3)$, *and an estimate of the form*

$$\|X\|_{C^{2,\mu}(\Omega')} \le C(\mu, \inf K, \|K\|_{C^\mu}, \sup |X|, \operatorname{dist}(\Omega', \partial\Omega)) \tag{10.25}$$

holds for all $\Omega' \subset\subset \Omega$.

Proof: Since the third component of X, $z(x,y)$, is a convex or concave function, one easily estimates for $x \in \Omega' \subset\subset \Omega$:

$$|Dz(x,y)| \le \frac{2\sup|z|}{\operatorname{dist}(\Omega',\partial\Omega)} . \quad \square \tag{10.26}$$

10.3. The Darboux equation and the regularity of locally convex surfaces

Suppose that Σ is a regular surface with radius vector $X \in C^{1,1}(\Omega,\mathbb{R}^3)$ and let

$$\rho = \rho(u^1,u^2) = X \cdot X_0 \tag{10.27}$$

with a fixed unit vector X_0. The Gauß equations (10.6) then imply that

$$D_{ij}\rho = \Gamma^k_{ij} D_k\rho + h_{ij}\nu \cdot X_0 , \tag{10.28}$$

hence

$$\begin{aligned}
\det\left[D_{ij}\rho - \Gamma^k_{ij}D_k\rho\right] &= h\,(\nu\cdot X_0)^2 \\
&= K\,(D_1X\ D_2X\ X_0)^2 \\
&= K\det\begin{bmatrix} |D_1X|^2 & D_1X\cdot D_2X & D_1X\cdot X_0 \\ D_2X\cdot D_1X & |D_2X|^2 & D_2X\cdot X_0 \\ X_0\cdot D_1X & X_0\cdot D_2X & |X_0|^2 \end{bmatrix} \\
&= K\left[g - \frac{g^{ij}}{g} D_i\rho D_j\rho\right] .
\end{aligned} \tag{10.29}$$

(10.29) is the *Darboux equation* which is elliptic if

$$K > 0 \quad \text{and} \quad \nu\cdot X_0 \neq 0 , \tag{10.30}$$

i.e., if Σ is a convex graph over a plane perpendicular to X_0.

Definition 10.3.1. A regular surface Σ is *locally convex* if the Gauß curvature K is positive.

Theorem 10.3.2. *Suppose that* Σ *is a locally convex surface with radius vector* $X \in C^{1,1}(\Omega,\mathbb{R}^3)$ *and such that* $I_\Sigma \in C^{2,\mu}(\Omega,\mathbb{R}^3)$ *for some* μ, $0<\mu<1$, *i.e,* $g_{ij} \in C^{2,\mu}(\Omega)$ *for* $i,j = 1,2$. *Then* $X \in C^{2,\mu}_{loc}(\Omega,\mathbb{R}^3)$.

Proof: For $(u_0^1,u_0^2) \in \Omega$ let

$$X_0 = \nu(u_0^1,u_0^2) .$$

The Darboux equation (10.29) is then elliptic in a neighborhood $\mathscr{N}$ of (u_0^1, u_0^2). Theorem 3.4.3 yields the regularity $\rho \in C^{2,\mu}_{loc}(\mathscr{N})$. To translate this into the regularity $X \in C^{2,\mu}_{loc}(\mathscr{N}, \mathbb{R}^3)$, consider the three 3×3–systems

$$X_0 \cdot D_{ij}X = D_{ij}\rho,$$

$$D_kX \cdot D_{ij}X = \frac{1}{2}(D_jg_{ik} + D_ig_{jk} - D_kg_{ij})$$

(which can easily be derived from the Gauß equations). By (10.29), the determinant of the coefficient matrix is

$$X_0\, D_1X\, D_2X = \sqrt{g - \frac{g^{ij}}{g} D_i\rho\, D_j\rho} \neq 0,$$

and the statement follows from Cramer's rule. □

10.4. Conjugate isothermal parameters and the Darboux system

Proposition 10.4.1. *Suppose that* Σ *is a locally convex surface with radius vector* $X \in C^{2,\mu}(\Omega, \mathbb{R}^3)$ *for some* μ, $0 < \mu < 1$, *and let* $\bar{B}_R = \bar{B}_R(u_0^1, u_0^2) \subset \Omega$. *Then there exists a homeomorphism* $(u^1, u^2) = (u^1(x_1, x_2), u^2(x_1, x_2))$ *from* $\bar{B} = \{x_1^2 + x_2^2 \leq 1\}$ *onto* $\bar{B}_R$ *of class* $C^{1,\mu}_{loc}(B)$ *with* $u^1(0) = u_0^1$, $u^2(0) = u_0^2$ *and*

$$Du^1 \wedge Du^2 = D_1u^1 D_2u^2 - D_2u^1 D_1u^2 \neq 0 \tag{10.31}$$

such that the following conformality relations hold:

$$\sqrt{gK}\, h^{ij} = \frac{Du^i \cdot Du^j}{Du^1 \wedge Du^2}, \tag{10.32}$$

i. e.,

$$\frac{h_{22}}{\sqrt{gK}} = \frac{|Du^1|^2}{Du^1 \wedge Du^2}, \tag{10.33}$$

$$-\frac{h_{12}}{\sqrt{gK}} = \frac{Du^1 \cdot Du^2}{Du^1 \wedge Du^2}, \tag{10.34}$$

$$\frac{h_{11}}{\sqrt{gK}} = \frac{|Du^2|^2}{Du^1 \wedge Du^2}. \tag{10.35}$$

Furthermore (u^1, u^2) *satisfies the Darboux system*

$$\Delta_K u^k = D_\alpha(\sqrt{K} D_\alpha u^k) + \sqrt{K}\Gamma^k_{ij} Du^i \cdot Du^j = 0 \qquad (k = 1, 2). \tag{10.36}$$

<u>Proof:</u> $X \in C^{2,\mu}(\Omega,\mathbb{R}^3)$ according to Theorem 10.3.2. Assume first that $X \in C^3(\Omega,\mathbb{R}^3)$, so that $II_\Sigma \in C^1(\Omega,\mathbb{R}^3)$, and consider the differential form

$$\begin{aligned} ds^2 &= \frac{1}{\sqrt{g}} II_\Sigma \\ &= \frac{h_{ij}}{\sqrt{g}} du^i du^j . \end{aligned}$$

Theorem 6.3.2 yields the existence of the parameters (x_1, x_2) which satisfy the conformality relations (10.32). According to (6.52, 53),

$$D_\alpha(\sqrt{K} D_\alpha u^1) = \left[D_1\left[\frac{h_{22}}{\sqrt{g}}\right] - D_2\left[\frac{h_{12}}{\sqrt{g}}\right]\right] Du^1 \wedge Du^2,$$

$$D_\alpha(\sqrt{K} D_\alpha u^2) = \left[D_2\left[\frac{h_{11}}{\sqrt{g}}\right] - D_1\left[\frac{h_{12}}{\sqrt{g}}\right]\right] Du^1 \wedge Du^2.$$

By invoking the Codazzi–Mainardi equations (10.12) and (10.10), one computes

$$\begin{aligned} D_1\left[\frac{h_{22}}{\sqrt{g}}\right] - D_2\left[\frac{h_{12}}{\sqrt{g}}\right] &= \frac{1}{\sqrt{g}}(\Gamma^\ell_{12} h_{2\ell} - \Gamma^\ell_{22} h_{1\ell} - h_{22}(\Gamma^1_{11} + \Gamma^2_{21}) + h_{12}(\Gamma^1_{12} + \Gamma^2_{22})) \\ &= \frac{1}{\sqrt{g}}(-\Gamma^1_{11} h_{22} + 2\Gamma^1_{12} h_{12} - \Gamma^1_{22} h_{11}) \\ &= -\sqrt{K}\Gamma^1_{ij} \frac{Du^i \cdot Du^j}{Du^1 \wedge Du^2}, \end{aligned}$$

and

$$\begin{aligned} D_2\left[\frac{h_{11}}{\sqrt{g}}\right] - D_1\left[\frac{h_{12}}{\sqrt{g}}\right] &= \frac{1}{\sqrt{g}}(\Gamma^\ell_{21} h_{1\ell} - \Gamma^\ell_{11} h_{2\ell} - h_{11}(\Gamma^1_{12} + \Gamma^2_{22}) + h_{12}(\Gamma^1_{11} + \Gamma^2_{21})) \\ &= \frac{1}{\sqrt{g}}(-\Gamma^2_{11} h_{22} + \Gamma^2_{21} h_{12} - \Gamma^2_{22} h_{11}) \\ &= -\sqrt{K}\Gamma^2_{ij} \frac{Du^i \cdot Du^j}{Du^1 \wedge Du^2}. \end{aligned}$$

The statement remains true if $X \in C^{2,\mu}(\Omega,\mathbb{R}^3)$, $I_\Sigma \in C^{2,\mu}(\Omega,\mathbb{R}^3)$. This is seen by essentially repeating the approximation argument in the proof of Theorem 6.3.2: By the theorema egregium we have $K \in C^\mu(\Omega)$, and hence there are $C^{1,\mu}_{loc}$–estimates of the form (6.60, 61) for the approximating mappings $\{(u^1,u^2)^{(n)}\}_{n=1}^\infty$. Such estimates also hold true for the inverses $\{(x_1,x_2)^{(n)}\}_{n=1}^\infty$ because of the integrability conditions (6.62), i.e., because of the system

$$D_j(\sqrt{h}\, h^{ij} D_i x_k^{(n)}) = 0$$

which has Hölder continuous coefficients. □

Remark 10.4.2. The parameters (x_1, x_2) are called *conjugate isothermal* because they are obtained by mapping $(\Omega, \lambda II_\Sigma)$ conformally onto the unit disc $(B, dx_1^2 + dx_2^2)$.

10.5. A priori estimates for locally convex surfaces

Lemma 10.5.1. *Let* Σ *be a locally convex surface with radius vector* $X \in C^{2,\mu}(\Omega, \mathbb{R}^3)$. *Suppose that*

$$|g_{ij}| \leq a, \tag{10.37}$$

$$g,\ K \geq \frac{1}{c}. \tag{10.38}$$

Then the mapping $(u^1, u^2) = (u^1(x_1, x_2), u^2(x_1, x_2))$, $(x_1, x_2) \in B$, *from Proposition* 10.4.1 *satisfies the estimate*

$$\int_B (|Du^1|^2 + |Du^2|^2)\, dx_1\, dx_2 \leq C(a, c) \int_\Sigma |H|\, d\sigma. \tag{10.39}$$

Proof: The mean curvature H of Σ can be estimated by the conformality relations (10.32) and by (3.5),

$$\begin{aligned}
|H| &= \left|\frac{h_{ij} g^{ij}}{2}\right| \\
&= \left|\frac{h}{2g} g_{ij} h^{ij}\right| \\
&= \frac{1}{2}\sqrt{\frac{K}{g}}\, g_{ij} \frac{Du^i \cdot Du^j}{|Du^1 \wedge Du^2|} \\
&\geq \frac{1}{2}\sqrt{\frac{K}{g}}\, \frac{g}{2a}\, \frac{|Du^1|^2 + |Du^2|^2}{|Du^1 \wedge Du^2|} \\
&\geq \frac{1}{4ac}\, \frac{|Du^1|^2 + |Du^2|^2}{|Du^1 \wedge Du^2|}
\end{aligned}$$

Therefore

$$\begin{aligned}
\int_B (|Du^1|^2 + |Du^2|^2)\, dx_1\, dx_2 &\leq 4ac \int_{B_R(u_0^1, u_0^2)} |H|\, du^1\, du^2 \\
&\leq 4a\sqrt{c} \int_\Sigma |H|\, d\sigma. \quad \square
\end{aligned}$$

<u>Theorem 10.5.2.</u> *Let* Σ *be a locally convex surface with radius vector* $X \in C^{1,1}(\Omega,\mathbb{R}^3)$ *and* $I_\Sigma \in C^{2,\mu}(\Omega,\mathbb{R}^3)$ *for some* μ, $0<\mu<1$. *Suppose that*

(10.40) $$\|g_{ij}\|_{C^{2,\mu}(\Omega)} \le a,$$

(10.41) $$g,\ K \ge \frac{1}{c},$$

(10.42) $$\int_\Sigma |H|\, d\sigma \le M.$$

Then $X \in C^{2,\mu}(\Omega,\mathbb{R}^3)$ *and for* $\Omega' \subset\subset \Omega$ *there is an estimate of the form*

(10.43) $$\|D^2X\|_{C^\mu(\Omega')} \le C,$$

where the constant C *only depends on* μ, a, c, M, $\mathrm{dist}(\Omega',\partial\Omega)$.

<u>Proof:</u> Let $\bar{B}_R(u_0^1,u_0^2) \subset \Omega$. Consider the homeomorphism $(u^1,u^2) = (u^1(x_1,x_2), u^2(x_1,x_2))$ from $\bar{B}$ onto $\bar{B}_R$ from Proposition 10.4.1. (u^1,u^2) is of class $C^{1,\mu}_{loc}(B)$ and, by Lemma 10.5.1, has finite Dirichlet integral. Furthermore

(10.44) $$\sqrt{gK}\,h^{ij} \doteq \frac{Du^i \cdot Du^j}{Du^1 \wedge \Delta u^2},$$

(10.45) $$\Delta_K u^k = 0.$$

Suppose now w.l.o.g. that $B_R(u_0^1,u_0^2) = B = \{(u^1)^2+(u^2)^2<1\}$ (otherwise consider the mapping $\frac{1}{R}(u^1(x_1,x_2)-u_0^1, u^2(x_1,x_2)-u_0^2)$). Theorem 8.3.2 can then be applied to the system (10.45) to give the following estimates in any disc $B_\rho = \{x_1^2+x_2^2<\rho^2\}$, $0<\rho<1$:

(10.46) $$\|u^1,u^2\|_{C^{1,\mu}(B_\rho)} \le C(\ldots,\rho),$$

(10.47) $$Du^1 \wedge Du^2 \ge c(\ldots,\rho) > 0.$$

By taking $\rho=\frac{1}{2}$, and then also taking into account that we assumed that $B_R(u_0^1,u_0^2)=B$, the relations (10.44) yield the estimates

$$|h_{ij}(u_0^1,u_0^2)| \le C(\ldots,R),$$

from which, in Ω',

$$|h_{ij}| \le C(\ldots,\mathrm{dist}(\Omega',\partial\Omega)).$$

Furthermore the functions $h_{ij}(u^1(x_1,x_2),u^2(x_1,x_2))$ satisfy Hölder estimates of the form

$$[h_{ij}]_{\mu}^{B_\rho} \le C(\dots,\rho,R)$$

in each disc $B_\rho = \{x_1^2+x_2^2 < \rho^2\}$. In order to translate this into estimates for $h_{ij}(u^1,u^2)$, note that the estimate (10.39) for the Dirichlet integral of (u^1,u^2) implies by the Courant–Lebesgue lemma, Lemma 1.6.3, that there exists a $\rho = \rho(a,c,M,R)$, $0<\rho<1$, such that $(x_1,x_2) \in B_\rho$ if $(u^1,u^2) \in B_{R/2}(u_0^1,u_0^2)$. Since

$$x_k = \int_0^1 \{D_1x_k(u_0^1+\tau(u^1-u_0^1),u_0^2+\tau(u^2-u_0^2))(u^1-u_0^1) + D_2x_k(\dots)(u^2-u_0^2)\}\,d\tau,$$

the estimates (10.46,47) yield a dilation estimate of the form

$$x_1^2 + x_2^2 \le C(\dots,R)((u^1-u_0^1)^2 + (u^2-u_0^2)^2)$$

if $(u^1,u^2) \in B_{R/2}(u_0^1,u_0^2)$. Therefore

$$|h_{ij}(u^1,u^2) - h_{ij}(u_0^1,u_0^2)| \le C(\dots,R)((u^1-u_0^1)^2 + (u^2-u_0^2)^2)^{\mu/2},$$

from which

$$[h_{ij}]_{\mu}^{\Omega'} \le C(\dots,\mathrm{dist}(\Omega',\partial\Omega)).$$

A priori estimates for the second derivatives of the radius vector X follow now from the Gauß equations (10.6) as required. □

REFERENCES

Adams, R.A.
[AD] Sobolev spaces. New York–San Francisco–London: Academic Press 1975.

Ahlfors, L.
[AF] Lectures on quasiconformal mappings. Princeton–Toronto–New York–London: Van Nostrand 1966.

Ahlfors, L. and L. Bers
[AB] Riemann's mapping theorem for variable metrics. Ann. Math. (2) **72** (1960), 385–404.

Alexandrow, A.D.
[AL] Die innere Geometrie der konvexen Flächen. Berlin: Akademie–Verlag 1955.

Amano, K.
[AM] The Dirichlet problem for degenerate elliptic Darboux equation. Centre Math. Anal. Preprint R38–87. Canberra: Australian Nat. Univ. 1987.

Aubin, T.
[AU] Nonlinear analysis on manifolds. Monge–Ampère equations. New York–Heidelberg–Berlin: Springer–Verlag 1982.

Bakel'man, I. Ya.
[BM] Generalized solutions of the Monge–Ampère equations. Dokl. Akad. Nauk SSSR **114** (1957), 1143–1145.

Berg, P.W.
[BG] On univalent mappings by solutions of linear elliptic partial differential equations. Trans. Amer. Math. Soc. **84** (1957), 310–318.

Bers, L.
[BS 1] Partial differential equations and generalized analytic functions. Proc. Nat. Acad. Sci. U.S.A. **37** (1951), 42–47.
[BS 2] Theory of pseudo–analytic functions. Lecture Notes. New York: New York Univ. 1953.
[BS 3] Univalent solutions of linear elliptic systems. Comm. Pure Appl. Math. **6** (1953), 513–526.
[BS 4] Local behavior of solutions of general linear elliptic equations. Comm. Pure. Appl. Math. **8** (1955), 473–496.
[BS 5] An outline of the theory of pseudoanalytic functions. Bull. Amer. Math. Soc. **62** (1956), 291–331.

Bers, L. and L. Nirenberg
[BN] On a representation theorem for linear elliptic systems with discontinuous coefficients and its applications. In: Convegno Internazionale sulle Equazioni Lineari alle Derivate Parziali (Triest 1954), pp. 111–140. Rome: Edizioni Cremonese 1955.

Campanato, S.
[CA 1] Proprietà di Hölderianità di alcune classi di funzioni. Ann. Scuola Norm. Sup. Pisa **17** (1963), 175–188.
[CA 2] Proprietà di una famiglia di spazi funzionali. Ann. Scuola Norm. Sup. Pisa **18** (1964), 137–160.
[CA 3] Equazioni ellittiche del II° ordine e spazi $\mathscr{L}^{(2,\lambda)}$. Ann. Mat. Pura Appl. (4) **69** (1965), 321–381.
[CA 4] Sistemi ellittici in forma divergenza. Regolarità all'interno. Quaderni. Pisa: Scuola Normale Superiore 1980.

Carleman, T.
[CM] Sur les systèmes linéaires aux dérivées partielles du premier ordre à deux variables. C. R. Acad. Sci. Paris (1) **197** (1933), 471–474.

Courant, R.
[CO] Dirichlet's principle, conformal mapping, and minimal surfaces. New York–London: Interscience 1950.

Courant, R. and Hilbert, D.
[CH] Methoden der mathematischen Physik II. 2nd ed., Berlin–Heidelberg–New York: Springer–Verlag 1968.

Calderón, H.P. and A. Zygmund
[CZ] On the existence of certain singular integrals. Acta Math. **88** (1952), 85–139.

Darboux, G.
[DB] Leçons sur la théorie générale des surfaces et les applications géométriques du calcul infinitésimal I, III. Paris: Gauthier–Villars 1887; 1894.

Delanoë, P.
[DL 1] Équations de Monge–Ampère en dimension deux. C. R. Acad. Sci. Paris **294** (1982), 693–696.
[DL 2] Réalisations globalement régulières de disques strictement convexes dans les espaces d'Euclide et de Minkowski par la méthode de Weingarten. Ann. Scient. Ec. Norm. Sup. (4) **21** (1988), 637–652.

Efimov, N.W.
[EF] Flächenverbiegung im Großen. Mit einem Nachtrag von E. Rembs und K.P. Grotemeyer. Berlin: Akademie–Verlag 1957.

Evans, L.C. and R.F. Gariepy
[EG] Lecture notes on measure theory and fine properties of functions. Kentucky EPSCoR Preprint Series. Lexington: Univ. of Kentucky 1987.

Friedrichs, K. and H. Lewy
[FL 1] Über die Eindeutigkeit und das Abhängigkeitsgebiet der Lösungen beim Anfangswertproblem linearer hyperbolischer Differentialgleichungen. Math. Ann. **98** (1928), 192–204.
[FL 2] Das Anfangswertproblem einer beliebigen hyperbolischen Differentialgleichung beliebiger Ordnung in zwei Variablen. Existenz, Eindeutigkeit und Abhängigkeitsgebiet. Math. Ann. **99** (1928) 200–221.

Giaquinta, M.
[GI] Multiple integrals in the calculus of variations and nonlinear elliptic systems. Ann. Math. Studies **105**. Princeton, N.J.: Princeton Univ. Press 1983.

Gilbarg, D. and N.S. Trudinger
[GT] Elliptic partial differential equations of second order. 2nd ed., Berlin–Heidelberg–New York–Tokyo: Springer–Verlag 1983.

Gillis, P.
[GL] Intégrales doubles du calcul des variations. Acad. Roy. Belgique, Bull. Cl. Sci. (5) **36** (1950), 403–412.

Goursat, E.
[GO] Cours d'analyse mathématique. 4th ed., Paris: Gauthier–Villars 1927.

Hadamard, J.
[HA 1] Sur les charactéristiques des systèmes aux dérivées partielles. Bull. Soc. Math. France **34** (1906), 48–52.
[HA 2] Le problème de Cauchy et les équations aux dérivées partielles linéaires hyperboliques. Paris: Hermann 1932.

Hartman, P. and A. Wintner
[HW 1] On the asymptotic curves of a surface. Amer. J. Math. **73** (1951), 149–172.
[HW 2] On the local behavior of solutions of non–parabolic partial differential equations. Amer. J. Math. **75** (1953), 449–476.
[HW 3] On elliptic Monge–Ampère equations. Amer. J. Math. **75** (1953), 611–620.

Heinz, E.

[H 1] Über gewisse elliptische Systeme von Differentialgleichugen zweiter Ordnung mit Anwendung auf die Monge–Ampèresche Gleichung. Math. Ann. **131** (1956), 411–428.

[H 2] On certain nonlinear elliptic differential equations and univalent mappings. J. Analyse Math. **5** (1956/57), 197–272.

[H 3] On elliptic Monge–Ampère equations and Weyl's embedding problem. J. Analyse Math. **7** (1959), 1–52.

[H 4] On one–to–one harmonic mappings. Pacific J. Math. **9** (1959), 101–105.

[H 5] Über die Differentialungleichung $0 < \alpha \le rt - s^2 \le \beta < \infty$. Math. Z. **72** (1959), 107–126.

[H 6] Neue a–priori–Abschätzungen für den Ortsvektor einer Fläche positiver Gaußscher Krümmung durch ihr Linienelement. Math. Z. **74** (1960), 129–157.

[H 7] Interior estimates for solutions of elliptic Monge–Ampère equations. In: Partial Differential Equations. Proc. Sympos. Pure Math., Vol. 4 (Berkeley, CA 1960), pp. 149–155. Providence, R.I.: Amer. Math. Soc. 1961.

[H 8] On Weyl's embedding problem. J. Math. Mech. **11** (1962), 421–454.

[H 9] Existence theorems for one–to–one mappings associated with elliptic systems of second order I, II. J. Analyse Math. **15** (1965), 325–352; **17** (1966), 145–184.

[H 10] A–priori–Abschätzungen für isometrische Einbettungen zweidimensionaler Riemannscher Mannigfaltigkeiten in drei dimensionale Riemannsche Räume. Math. Z. **100** (1967), 1–16.

[H 11] Über das Nichtverschwinden der Funktionaldeterminante bei einer Klasse eineindeu–tiger Abbildungen. Math. Z. **105** (1968), 87–89.

[H 12] Zur Abschätzung der Funktionaldeterminante bei einer Klasse topologischer Abbil–dungen. Nachr. Akad. Wiss. Göttingen, II. Math.–Phys. Kl. **1968**, 183–197.

Jörgens, K.

[JÖ 1] Über die Lösungen der Differentialgleichung $rt - s^2 = 1$. Math. Ann. **127** (1954), 130–134.

[JÖ 2] Harmonische Abbildungen und die Differentialgleichung $rt - s^2 = 1$. Math. Ann. **129** (1955), 330–344.

Jost, J.

[JT 1] Univalency of harmonic mappings between surfaces. J. Reine Angew. Math. **324** (1981), 141–153.

[JT 2] Harmonic mappings between Riemannian manifolds. Proc. Centre Math. Anal. **4**. Canberra: Australian Nat. Univ. 1983.

Jost, J. and R. Schoen

[JS] On the existence of harmonic diffeomorphisms between surfaces. Invent. Math. **66** (1982), 353–359.

Kneser, H.

[KN] Lösung der Aufgabe 41. Jahresber. Deutschen Math. Vereinigung **35** (1926), 123–124.

Lewy, H.

[L 1] Über das Anfangswertproblem bei einer hyperbolischen nichtlinearen partiellen Differentialgleichung zweiter Ordnung mit zwei unabhängigen Veränderlichen. Math. Ann. **98** (1928), 179–191.

[L 2] Neuer Beweis des analytischen Charakters der Lösungen elliptischer Differential–gleichungen. Math. Ann. **101** (1929), 609–619; **107** (1929), 804.

[L 3] A priori limitations for solutions of Monge–Ampère equations I, II. Trans. Amer. Math. Soc. **37** (1935), 417–434; **41** (1937), 365–374.

[L 4] On the non–vanishing of the Jacobian in certain one–to–one mappings. Bull. Amer. Math. Soc. **42** (1936), 689–692.

[L 6] On the existence of a closed convex surface realizing a given Riemannian metric. Proc. Nat. Acad. Sci. U.S.A. **24** (1938), 104–106.

[L 7] A property of spherical harmonics. Amer. J. Math. **60** (1938), 555–560.

[L 8] On the non–vanishing of the Jacobian of a homeomorphism by harmonic gradients. Ann. Math. (2) **88** (1968), 518–529.

[L 9] About the Hessian of a spherical harmonic. Amer. J. Math. **91** (1969), 505–507.

Meyers, N.G.
[ME] Mean oscillation over cubes and Hölder continuity. Proc. Amer. Math. Soc. **15** (1964), 717–721.

Meyers, N.G. and Serrin J.
[MS] H = W. Proc. Nat. Acad. Sci. U.S.A. **51** (1964), 1055–1056.

Morrey, C.B., Jr.
[MO 1] Second order elliptic systems of differential equations. In: Ann. Math. Studies **33**, pp. 101–159. Princeton, N.J.: Princeton Univ. Press 1954.
[MO 2] Multiple integrals in the calculus of variations. New York: Springer–Verlag 1966.

Nevanlinna, R.
[NE] Uniformisierung. Berlin–Göttingen–Heidelberg: Springer–Verlag 1953.

Nikolaiev, I.G. and C.Z. Shefel'
[NIS 1] Smoothness of convex surfaces on the basis of differential properties of quasi–conformal mappings. Soviet Math. Dokl. **26** (1982), 599–602.
[NIS 2] Convex surfaces with positive bounded specific curvature and a priori estimates for Monge–Ampère equations. Siberian Math. J. **26** (1985), 572–586.

Nirenberg, L.
[NI 1] On nonlinear elliptic partial differential equations and Hölder continuity. Comm. Pure Appl. Math. **6** (1953), 103–156.
[NI 2] The Weyl and Minkowski problems in differential geometry in the large. Comm. Pure Appl. Math. **6** (1953), 337–394.

Nitsche, J.C.C.
[NT 1] Elementary proof of Bernstein's theorem on minimal surfaces. Ann. Math. **66** (1957), 543–544.
[NT 2] On harmonic mappings. Proc. Amer. Math. Soc. **9** (1958), 268–271.
[NT 3] Vorlesungen über Minimalflächen. Berlin–Heidelberg–New York: Springer–Verlag 1975.

Pogorelov, A.V.
[PO 1] Die Verbiegung konvexer Flächen. Berlin: Akademie–Verlag 1957.
[PO 2] Monge–Ampère equations of elliptic type. Groningen: Noordhoff 1964.
[PO 3] Some results on surface theory in the large. Advances Math. **1** (1964), 191–264.
[PO 4] Extrinsic geometry of convex surfaces. Providence, R.I.: Amer. Math. Soc. 1973.

Pólya, G. and Szegö
[PS] Aufgaben und Lehrsätze aus der Analysis. 2nd ed., Berlin–Göttingen–Heidelberg: Springer–Verlag 1954.

Radó, T.
[RA] Aufgabe 41. Jahresber. Deutschen Math. Vereinigung **35** (1926), 123–124.

Rellich, F.
[RE 1] Über die Reduktion gewisser ausgearteter Systeme von partiellen Differential–gleichungen. Math. Ann. **109** (1934), 714–745.
[RE 2] Die Bestimmung einer Fläche durch ihre Gaußsche Krümmung. Math. Z. **43** (1938), 618–627.

Renelt, H.
[RT] Elliptic systems and quasiconformal mappings. Chichester–New York–Brisbane–Toronto–Singapore: Wiley–Interscience 1988.

Sabitov, I. Kh.
[SB 1] The regularity of convex regions with a metric that is regular in the Hölder classes. Siberian Math. J. **17** (1976), 681–687.
[SB 2] A successive approximation scheme for the immersions of two–dimensional metrics into E^3. Siberian Math. J. **19** (1978), 957–973.

Sabitov, I. Kh. and S.Z. Shefel'
[SBS] The connections between the order of smoothness of a surface and its metric. Siberian Math. J. **17** (1976), 687–694.

Safonov, M.V.
[SF] On the classical solution of Bellman's elliptic equation. Soviet Math. Dokl. **30** (1984), 482–485.

Schauder, J.
[SC] Über lineare elliptische Differentialgleichungen zweiter Ordung. Math. Z. **38** (1934), 257–282.

Schiffer, M. and D.C. Spencer
[SHS] Functionals of finite Riemann surfaces. Princeton, N.J.: Princeton Univ. Press 1954.

Schulz, F.
[SZ 1] Über elliptische Monge–Ampèresche Differentialgleichungen mit einer Bemerkung zum Weylschen Einbettungsproblem. Nachr. Akad. Wiss. Göttingen, II. Math.–Phys. Kl. **1981**, 93–108.
[SZ 2] Über die Differentialgleichung $rt - s^2 = f$ und das Weylsche Einbettungsproblem. Math. Z. **179** (1982), 1–10.
[SZ 3] A priori estimates for solutions of Monge–Ampère equations. Arch. Rational Mech. Anal. **89** (1985), 123–133.
[SZ 4] Boundary estimates for solutions of Monge–Ampère equations in the plane. Ann. Scuola Norm. Sup. Pisa, Cl. Sci. (4) **11** (1984), 431–440.
[SZ 5] Univalent solutions of elliptic systems of Heinz–Lewy type. Ann. Inst. Henri Poincaré, Anal. Non Linéaire **6** (1989), 347–361.
[SZ 6] Second derivative estimates for solutions of two–dimensional Monge–Ampère equations. Proc. Amer. Math. Soc. (to appear).
[SZ 7] Regularity of locally convex surfaces. Bull. Austr. Math. Soc. (to appear).

Schulz, F. and L.–Y. Liao
[SL] Regularity of solutions of two–dimensional Monge–Ampère equations. Trans. Amer. Math. Soc. **307** (1988), 271–277.

Segre, B.
[SG] Questioni di realità sulle forme armoniche e sulle loro hessiane I, II. Rend. Accad. Naz. Lincei, Cl. Sci. Fis. Mat. Natur. (8) **15** (1953), 237–242; 339–344.

Sen'kin, E.P.
[SK] Bendings of convex surfaces. Itogi Nauki Tekhniki Problemy Geometrii **10** (1978), 193–224 [Russian].

Shefel', S.Z.
[SH 1] Smoothness of the solution to the Minkowski problem. Siberian Math. J. **18** (1977), 338–340.
[SH 2] Geometric properties of immersed manifolds. Siberian Math. J. **26** ((1985), 133–147.

Shikin, E.V.
[SN] Equations of isometric imbeddings in three–dimensional Euclidean space of two–dimensional manifolds of negative curvature. Math. Notes Acad. Sc. USSR **31** (1982), 305–312.

Shishkov, A.E.
[SV] On existence of solutions, regular up to the domain boundary, of the Dirichlet problem for strongly elliptic Monge–Ampère equations. Dopovidi Akad. Nauk. Ukrain. RSR, Ser. A, **40** (1978), 30–33 [Russian].

Skypnik, I.V. and A.E. Shishkov
[SHS] On the solution of the Dirichlet problem for the Monge–Ampère equations. Dopovidi Akad. Nauk Ukrain. RSR, Ser. A, **40** (1978), 215–218.

Spivak, M.
[SV] A comprehensive introduction to differential geometry IV, V. Wilmington, Del.: Publish or Perish 1975.

Talenti, G.
[TA] Some estimates of solutions to Monge–Ampère type equations in dimension two. Ann. Scuola Norm. Super. Pisa, Cl. Sci. (4) **8** (1981), 183–230.

Vekua, I.N.

[VE 1] Systems of differential equations of the first order of elliptic type and boundary value problems. Applications to the theory of shells. Mat. Sbornik **31** (1952), 217–314.

[VE 2] Generalized analytic functions. London–Paris–Frankfurt: Pergamon Press 1962.

Weyl, H.

[WE] Über die Bestimmung einer geschlossenen konvexen Fläche durch ihr Linienelement. In: Selecta Hermann Weyl, pp. 148–178. Basel–Stuttgart: Birkhäuser Verlag 1956.

NOTATION INDEX

SUBJECT INDEX

LECTURE NOTES IN MATHEMATICS

Edited by A. Dold, B. Eckmann and F. Takens

Some general remarks on the publication of monographs and seminars

In what follows all references to monographs, are applicable also to multiauthorship volumes such as seminar notes.

§1. Lecture Notes aim to report new developments - quickly, informally, and at a high level. Monograph manuscripts should be reasonably self-contained and rounded off. Thus they may, and often will, present not only results of the author but also related work by other people. Furthermore, the manuscripts should provide sufficient motivation, examples and applications. This clearly distinguishes Lecture Notes manuscripts from journal articles which normally are very concise. Articles intended for a journal but too long to be accepted by most journals, usually do not have this "lecture notes" character. For similar reasons it is unusual for Ph.D. theses to be accepted for the Lecture Notes series.

Experience has shown that English language manuscripts achieve a much wider distribution.

§2. Manuscripts or plans for Lecture Notes volumes should be submitted (preferably in duplicate) either to one of the series editors or to Springer- Verlag, Heidelberg. These proposals are then refereed. A final decision concerning publication can only be made on the basis of the complete manuscripts, but a preliminary decision can usually be based on partial information: a fairly detailed outline describing the planned contents of each chapter, and an indication of the estimated length, a bibliography, and one or two sample chapters - or a first draft of the manuscript. The editors will try to make the preliminary decision as definite as they can on the basis of the available information. We generally advise authors not to prepare the final master copy of their manuscript (cf. §4) beforehand.

§3. Final manuscripts should contain at least 100 pages of mathematical text and should include

- a table of contents;
- an informative introduction, perhaps with some historical remarks: it should be accessible to a reader not particularly familiar with the topic treated;
- a subject index: this is almost always genuinely helpful for the reader.

§4. Lecture Notes are printed by photo-offset from the master-copy delivered in camera-ready form by the authors. Springer-Verlag provides technical instructions for the preparation of manuscripts, for typewritten manuscripts special stationery, with the prescribed typing area outlined, is available on request. Careful preparation of the manuscripts will help keep production time short and ensure satisfactory appearance of the finished book. For manuscripts typed or typeset according to our instructions, Springer-Verlag will, if necessary, contribute towards the preparation costs at a fixed rate.

The actual production of a Lecture Notes volume takes 6-8 weeks.

§5. Authors receive a total of 50 free copies of their volume, but no royalties. They are entitled to purchase further copies of their book for their personal use at a discount of 33.3 %, other Springer mathematics books at a discount of 20 % directly from Springer-Verlag.

Commitment to publish is made by letter of intent rather than by signing a formal contract. Springer-Verlag secures the copyright for each volume.

<u>Addresses:</u>

Professor A. Dold, Mathematisches Institut, Universität Heidelberg,
Im Neuenheimer Feld 288, 6900 Heidelberg, Federal Republic of Germany

Professor B. Eckmann, Mathematik, ETH-Zentrum
8092 Zürich, Switzerland

Prof. F. Takens, Mathematisch Instituut, Rijksuniversiteit Groningen,
Postbus 800, 9700 AV Groningen, The Netherlands

Springer-Verlag, Mathematics Editorial, Tiergartenstr. 17,
6900 Heidelberg, Federal Republic of Germany, Tel.: (06221) 487-410

Springer-Verlag, Mathematics Editorial, 175 Fifth Avenue,
New York, New York 10010, USA, Tel.: (212) 460-1596

Vol. 1290: G. Wüstholz (Ed.), Diophantine Approximation and Transcendence Theory. Seminar, 1985. V, 243 pages. 1987.

Vol. 1291: C. Mœglin, M.-F. Vignéras, J.-L. Waldspurger, Correspondances de Howe sur un Corps p-adique. VII, 163 pages. 1987

Vol. 1292: J.T. Baldwin (Ed.), Classification Theory. Proceedings, 1985. VI, 500 pages. 1987.

Vol. 1293: W. Ebeling, The Monodromy Groups of Isolated Singularities of Complete Intersections. XIV, 153 pages. 1987.

Vol. 1294: M. Queffélec, Substitution Dynamical Systems – Spectral Analysis. XIII, 240 pages. 1987.

Vol. 1295: P. Lelong, P. Dolbeault, H. Skoda (Réd.), Séminaire d'Analyse P. Lelong – P. Dolbeault – H. Skoda. Seminar, 1985/1986. VII, 283 pages. 1987.

Vol. 1296: M.-P. Malliavin (Ed.), Séminaire d'Algèbre Paul Dubreil et Marie-Paule Malliavin. Proceedings, 1986. IV, 324 pages. 1987.

Vol. 1297: Zhu Y.-I., Guo B.-y. (Eds.), Numerical Methods for Partial Differential Equations. Proceedings. XI, 244 pages. 1987.

Vol. 1298: J. Aguadé, R. Kane (Eds.), Algebraic Topology, Barcelona 1986. Proceedings. X, 255 pages. 1987.

Vol. 1299: S. Watanabe, Yu.V. Prokhorov (Eds.), Probability Theory and Mathematical Statistics. Proceedings, 1986. VIII, 589 pages. 1988.

Vol. 1300: G.B. Seligman, Constructions of Lie Algebras and their Modules. VI, 190 pages. 1988.

Vol. 1301: N. Schappacher, Periods of Hecke Characters. XV, 160 pages. 1988.

Vol. 1302: M. Cwikel, J. Peetre, Y. Sagher, H. Wallin (Eds.), Function Spaces and Applications. Proceedings, 1986. VI, 445 pages. 1988.

Vol. 1303: L. Accardi, W. von Waldenfels (Eds.), Quantum Probability and Applications III. Proceedings, 1987. VI, 373 pages. 1988.

Vol. 1304: F.Q. Gouvêa, Arithmetic of *p*-adic Modular Forms. VIII, 121 pages. 1988.

Vol. 1305: D.S. Lubinsky, E.B. Saff, Strong Asymptotics for Extremal Polynomials Associated with Weights on $\mathbb{R}$. VII, 153 pages. 1988.

Vol. 1306: S.S. Chern (Ed.), Partial Differential Equations. Proceedings, 1986. VI, 294 pages. 1988.

Vol. 1307: T. Murai, A Real Variable Method for the Cauchy Transform, and Analytic Capacity. VIII, 133 pages. 1988.

Vol. 1308: P. Imkeller, Two-Parameter Martingales and Their Quadratic Variation. IV, 177 pages. 1988.

Vol. 1309: B. Fiedler, Global Bifurcation of Periodic Solutions with Symmetry. VIII, 144 pages. 1988.

Vol. 1310: O.A. Laudal, G. Pfister, Local Moduli and Singularities. V, 117 pages. 1988.

Vol. 1311: A. Holme, R. Speiser (Eds.), Algebraic Geometry, Sundance 1986. Proceedings. VI, 320 pages. 1988.

Vol. 1312: N.A. Shirokov, Analytic Functions Smooth up to the Boundary. III, 213 pages. 1988.

Vol. 1313: F. Colonius, Optimal Periodic Control. VI, 177 pages. 1988.

Vol. 1314: A. Futaki, Kähler-Einstein Metrics and Integral Invariants. IV, 140 pages. 1988.

Vol. 1315: R.A. McCoy, I. Ntantu, Topological Properties of Spaces of Continuous Functions. IV, 124 pages. 1988.

Vol. 1316: H. Korezlioglu, A.S. Ustunel (Eds.), Stochastic Analysis and Related Topics. Proceedings, 1986. V, 371 pages. 1988.

Vol. 1317: J. Lindenstrauss, V.D. Milman (Eds.), Geometric Aspects of Functional Analysis. Seminar, 1986–87. VII, 289 pages. 1988.

Vol. 1318: Y. Felix (Ed.), Algebraic Topology – Rational Homotopy. Proceedings, 1986. VIII, 245 pages. 1988

Vol. 1319: M. Vuorinen, Conformal Geometry and Quasiregular Mappings. XIX, 209 pages. 1988.

Vol. 1320: H. Jürgensen, G. Lallement, H.J. Weinert (Eds.), Semigroups, Theory and Applications. Proceedings, 1986. X, 416 pages. 1988.

Vol. 1321: J. Azéma, P.A. Meyer, M. Yor (Eds.), Séminaire de Probabilités XXII. Proceedings. IV, 600 pages. 1988.

Vol. 1322: M. Métivier, S. Watanabe (Eds.), Stochastic Analysis. Proceedings, 1987. VII, 197 pages. 1988.

Vol. 1323: D.R. Anderson, H.J. Munkholm, Boundedly Controlled Topology. XII, 309 pages. 1988.

Vol. 1324: F. Cardoso, D.G. de Figueiredo, R. Iório, O. Lopes (Eds.), Partial Differential Equations. Proceedings, 1986. VIII, 433 pages. 1988.

Vol. 1325: A. Truman, I.M. Davies (Eds.), Stochastic Mechanics and Stochastic Processes. Proceedings, 1986. V, 220 pages. 1988.

Vol. 1326: P.S. Landweber (Ed.), Elliptic Curves and Modular Forms in Algebraic Topology. Proceedings, 1986. V, 224 pages. 1988.

Vol. 1327: W. Bruns, U. Vetter, Determinantal Rings. VII,236 pages. 1988.

Vol. 1328: J.L. Bueso, P. Jara, B. Torrecillas (Eds.), Ring Theory. Proceedings, 1986. IX, 331 pages. 1988.

Vol. 1329: M. Alfaro, J.S. Dehesa, F.J. Marcellan, J.L. Rubio de Francia, J. Vinuesa (Eds.): Orthogonal Polynomials and their Applications. Proceedings, 1986. XV, 334 pages. 1988.

Vol. 1330: A. Ambrosetti, F. Gori, R. Lucchetti (Eds.), Mathematical Economics. Montecatini Terme 1986. Seminar. VII, 137 pages. 1988.

Vol. 1331: R. Bamón, R. Labarca, J. Palis Jr. (Eds.), Dynamical Systems, Valparaiso 1986. Proceedings. VI, 250 pages. 1988.

Vol. 1332: E. Odell, H. Rosenthal (Eds.), Functional Analysis. Proceedings, 1986–87. V, 202 pages. 1988.

Vol. 1333: A.S. Kechris, D.A. Martin, J.R. Steel (Eds.), Cabal Seminar 81–85. Proceedings, 1981–85. V, 224 pages. 1988.

Vol. 1334: Yu.G. Borisovich, Yu. E. Gliklikh (Eds.), Global Analysis – Studies and Applications III. V, 331 pages. 1988.

Vol. 1335: F. Guillén, V. Navarro Aznar, P. Pascual-Gainza, F. Puerta, Hyperrésolutions cubiques et descente cohomologique. XII, 192 pages. 1988.

Vol. 1336: B. Helffer, Semi-Classical Analysis for the Schrödinger Operator and Applications. V, 107 pages. 1988.

Vol. 1337: E. Sernesi (Ed.), Theory of Moduli. Seminar, 1985. VIII, 232 pages. 1988.

Vol. 1338: A.B. Mingarelli, S.G. Halvorsen, Non-Oscillation Domains of Differential Equations with Two Parameters. XI, 109 pages. 1988.

Vol. 1339: T. Sunada (Ed.), Geometry and Analysis of Manifolds. Procedings, 1987. IX, 277 pages. 1988.

Vol. 1340: S. Hildebrandt, D.S. Kinderlehrer, M. Miranda (Eds.), Calculus of Variations and Partial Differential Equations. Proceedings, 1986. IX, 301 pages. 1988.

Vol. 1341: M. Dauge, Elliptic Boundary Value Problems on Corner Domains. VIII, 259 pages. 1988.

Vol. 1342: J.C. Alexander (Ed.), Dynamical Systems. Proceedings, 1986–87. VIII, 726 pages. 1988.

Vol. 1343: H. Ulrich, Fixed Point Theory of Parametrized Equivariant Maps. VII, 147 pages. 1988.

Vol. 1344: J. Král, J. Lukeš, J. Netuka, J. Veselý (Eds.), Potential Theory – Surveys and Problems. Proceedings, 1987. VIII, 271 pages. 1988.

Vol. 1345: X. Gomez-Mont, J. Seade, A. Verjovski (Eds.), Holomorphic Dynamics. Proceedings, 1986. VII, 321 pages. 1988.

Vol. 1346: O. Ya. Viro (Ed.), Topology and Geometry – Rohlin Seminar. XI, 581 pages. 1988.

Vol. 1347: C. Preston, Iterates of Piecewise Monotone Mappings on an Interval. V, 166 pages. 1988.

Vol. 1348: F. Borceux (Ed.), Categorical Algebra and its Applications. Proceedings, 1987. VIII, 375 pages. 1988.

Vol. 1349: E. Novak, Deterministic and Stochastic Error Bounds in Numerical Analysis. V, 113 pages. 1988.

Vol. 1350: U. Koschorke (Ed.), Differential Topology. Proceedings, 1987. VI, 269 pages. 1988.

Vol. 1351: I. Laine, S. Rickman, T. Sorvali, (Eds.), Complex Analysis, Joensuu 1987. Proceedings. XV, 378 pages. 1988.

Vol. 1352: L.L. Avramov, K.B. Tchakerian (Eds.), Algebra – Some Current Trends. Proceedings, 1986. IX, 240 Seiten. 1988.

Vol. 1353: R.S. Palais, Ch.-l. Terng, Critical Point Theory and Submanifold Geometry. X, 272 pages. 1988.

Vol. 1354: A. Gómez, F. Guerra, M.A. Jiménez, G. López (Eds.), Approximation and Optimization. Proceedings, 1987. VI, 280 pages. 1988.

Vol. 1355: J. Bokowski, B. Sturmfels, Computational Synthetic Geometry. V, 168 pages. 1989.

Vol. 1356: H. Volkmer, Multiparameter Eigenvalue Problems and Expansion Theorems. VI, 157 pages. 1988.

Vol. 1357: S. Hildebrandt, R. Leis (Eds.), Partial Differential Equations and Calculus of Variations. VI, 423 pages. 1988.

Vol. 1358: D. Mumford, The Red Book of Varieties and Schemes. V, 309 pages. 1988.

Vol. 1359: P. Eymard, J.-P. Pier (Eds.), Harmonic Analysis. Proceedings, 1987. VIII, 287 pages. 1988.

Vol. 1360: G. Anderson, C. Greengard (Eds.), Vortex Methods. Proceedings, 1987. V, 141 pages. 1988.

Vol. 1361: T. tom Dieck (Ed.), Algebraic Topology and Transformation Groups. Proceedings, 1987. VI, 298 pages. 1988.

Vol. 1362: P. Diaconis, D. Elworthy, H. Föllmer, E. Nelson, G.C. Papanicolaou, S.R.S. Varadhan. École d'Été de Probabilités de Saint-Flour XV–XVII, 1985–87. Editor: P.L. Hennequin. V, 459 pages. 1988.

Vol. 1363: P.G. Casazza, T.J. Shura. Tsirelson's Space. VIII, 204 pages. 1988.

Vol. 1364: R.R. Phelps, Convex Functions, Monotone Operators and Differentiability. IX, 115 pages. 1989.

Vol. 1365: M. Giaquinta (Ed.), Topics in Calculus of Variations. Seminar, 1987. X, 196 pages. 1989.

Vol. 1366: N. Levitt, Grassmannians and Gauss Maps in PL-Topology. V, 203 pages. 1989.

Vol. 1367: M. Knebusch, Weakly Semialgebraic Spaces. XX, 376 pages. 1989.

Vol. 1368: R. Hübl, Traces of Differential Forms and Hochschild Homology. III, 111 pages. 1989.

Vol. 1369: B. Jiang, Ch.-K. Peng, Z. Hou (Eds.), Differential Geometry and Topology. Proceedings, 1986–87. VI, 366 pages. 1989.

Vol. 1370: G. Carlsson, R.L. Cohen, H.R. Miller, D.C. Ravenel (Eds.), Algebraic Topology. Proceedings, 1986. IX, 456 pages. 1989.

Vol. 1371: S. Glaz, Commutative Coherent Rings. XI, 347 pages. 1989.

Vol. 1372: J. Azéma, P.A. Meyer, M. Yor (Eds.), Séminaire de Probabilités XXIII. Proceedings. IV, 583 pages. 1989.

Vol. 1373: G. Benkart, J.M. Osborn (Eds.), Lie Algebras, Madison 1987. Proceedings. V, 145 pages. 1989.

Vol. 1374: R.C. Kirby, The Topology of 4-Manifolds. VI, 108 pages. 1989.

Vol. 1375: K. Kawakubo (Ed.), Transformation Groups. Proceedings, 1987. VIII, 394 pages, 1989.

Vol. 1376: J. Lindenstrauss, V.D. Milman (Eds.), Geometric Aspects of Functional Analysis. Seminar (GAFA) 1987–88. VII, 288 pages. 1989.

Vol. 1377: J.F. Pierce, Singularity Theory, Rod Theory, and Symmetry-Breaking Loads. IV, 177 pages. 1989.

Vol. 1378: R.S. Rumely, Capacity Theory on Algebraic Curves. III, 437 pages. 1989.

Vol. 1379: H. Heyer (Ed.), Probability Measures on Groups IX. Proceedings, 1988. VIII, 437 pages. 1989

Vol. 1380: H.P. Schlickewei, E. Wirsing (Eds.), Number Th 1987. Proceedings. V, 266 pages. 1989.

Vol. 1381: J.-O. Strömberg, A. Torchinsky. Weighted Hardy V, 193 pages. 1989.

Vol. 1382: H. Reiter, Metaplectic Groups and Segal Algebras pages. 1989.

Vol. 1383: D.V. Chudnovsky, G.V. Chudnovsky, H. Cohn, M.B. N (Eds.), Number Theory, New York 1985–88. Seminar. V, 256 page

Vol. 1384: J. Garcia-Cuerva (Ed.), Harmonic Analysis an Differential Equations. Proceedings, 1987. VII, 213 pages. 1989

Vol. 1385: A.M. Anile, Y. Choquet-Bruhat (Eds.), Relativistic Fl mics. Seminar, 1987. V, 308 pages. 1989.

Vol. 1386: A. Bellen, C.W. Gear, E. Russo (Eds.), Numerical M Ordinary Differential Equations. Proceedings, 1987. VII, 136 page

Vol. 1387: M. Petković, Iterative Methods for Simultaneous In Polynomial Zeros. X, 263 pages. 1989.

Vol. 1388: J. Shinoda, T.A. Slaman, T. Tugué (Eds.), Mathemat and Applications. Proceedings, 1987. V, 223 pages. 1989.

Vol. 1000: Second Edition. H. Hopf, Differential Geometry in the l 184 pages. 1989.

Vol. 1389: E. Ballico, C. Ciliberto (Eds.), Algebraic Curves and Geometry. Proceedings, 1988. V, 288 pages. 1989.

Vol. 1390: G. Da Prato, L. Tubaro (Eds.), Stochastic Partial D Equations and Applications II. Proceedings, 1988. VI, 258 page

Vol. 1391: S. Cambanis, A. Weron (Eds.), Probability Theory Spaces IV. Proceedings, 1987. VIII, 424 pages. 1989.

Vol. 1392: R. Silhol, Real Algebraic Surfaces. X, 215 pages. 198

Vol. 1393: N. Bouleau, D. Feyel, F. Hirsch, G. Mokobodz Séminaire de Théorie du Potentiel Paris, No. 9. Proceedings pages. 1989.

Vol. 1394: T.L. Gill, W.W. Zachary (Eds.), Nonlinear Semigrou Differential Equations and Attractors. Proceedings, 1987. IX, 2 1989.

Vol. 1395: K. Alladi (Ed.), Number Theory, Madras 1987. Procee 234 pages. 1989.

Vol. 1396: L. Accardi, W. von Waldenfels (Eds.), Quantum Proba Applications IV. Proceedings, 1987. VI, 355 pages. 1989.

Vol. 1397: P.R. Turner (Ed.), Numerical Analysis and Parallel Pr Seminar, 1987. VI, 264 pages. 1989.

Vol. 1398: A.C. Kim, B.H. Neumann (Eds.), Groups – Korea 1 ceedings. V, 189 pages. 1989.

Vol. 1399: W.-P. Barth, H. Lange (Eds.), Arithmetic of Complex M Proceedings, 1988. V, 171 pages. 1989.

Vol. 1400: U. Jannsen. Mixed Motives and Algebraic K-Theory. pages. 1990.

Vol. 1401: J. Steprāns, S. Watson (Eds.), Set Theory and its App Proceedings, 1987. V, 227 pages. 1989.

Vol. 1402: C. Carasso, P. Charrier, B. Hanouzet, J.-L. Jo Nonlinear Hyperbolic Problems. Proceedings, 1988. V, 249 page

Vol. 1403: B. Simeone (Ed.), Combinatorial Optimization. Semir V, 314 pages. 1989.

Vol. 1404: M.-P. Malliavin (Ed.), Séminaire d'Algèbre Paul Dubreil Paul Malliavin. Proceedings, 1987 – 1988. IV, 410 pages. 1989.

Vol. 1405: S. Dolecki (Ed.), Optimization. Proceedings, 198 pages. 1989.

Vol. 1406: L. Jacobsen (Ed.), Analytic Theory of Continued Frac Proceedings, 1988. VI, 142 pages. 1989.

Vol. 1407: W. Pohlers, Proof Theory. VI, 213 pages. 1989.

Vol. 1408: W. Lück, Transformation Groups and Algebraic K-Th 443 pages. 1989.

Vol. 1409: E. Hairer, Ch. Lubich, M. Roche. The Numerical So Differential-Algebraic Systems by Runge-Kutta Methods. VII, 13 1989.

Vol. 1410: F.J. Carreras, O. Gil-Medrano, A.M. Naveira (Eds.), Di Geometry. Proceedings, 1988. V, 308 pages. 1989.